MARINER
MISSION TO VENUS

$1.45

BY THE STAFF, Jet Propulsion Laboratory,
California Institute of Technology

Table of Contents

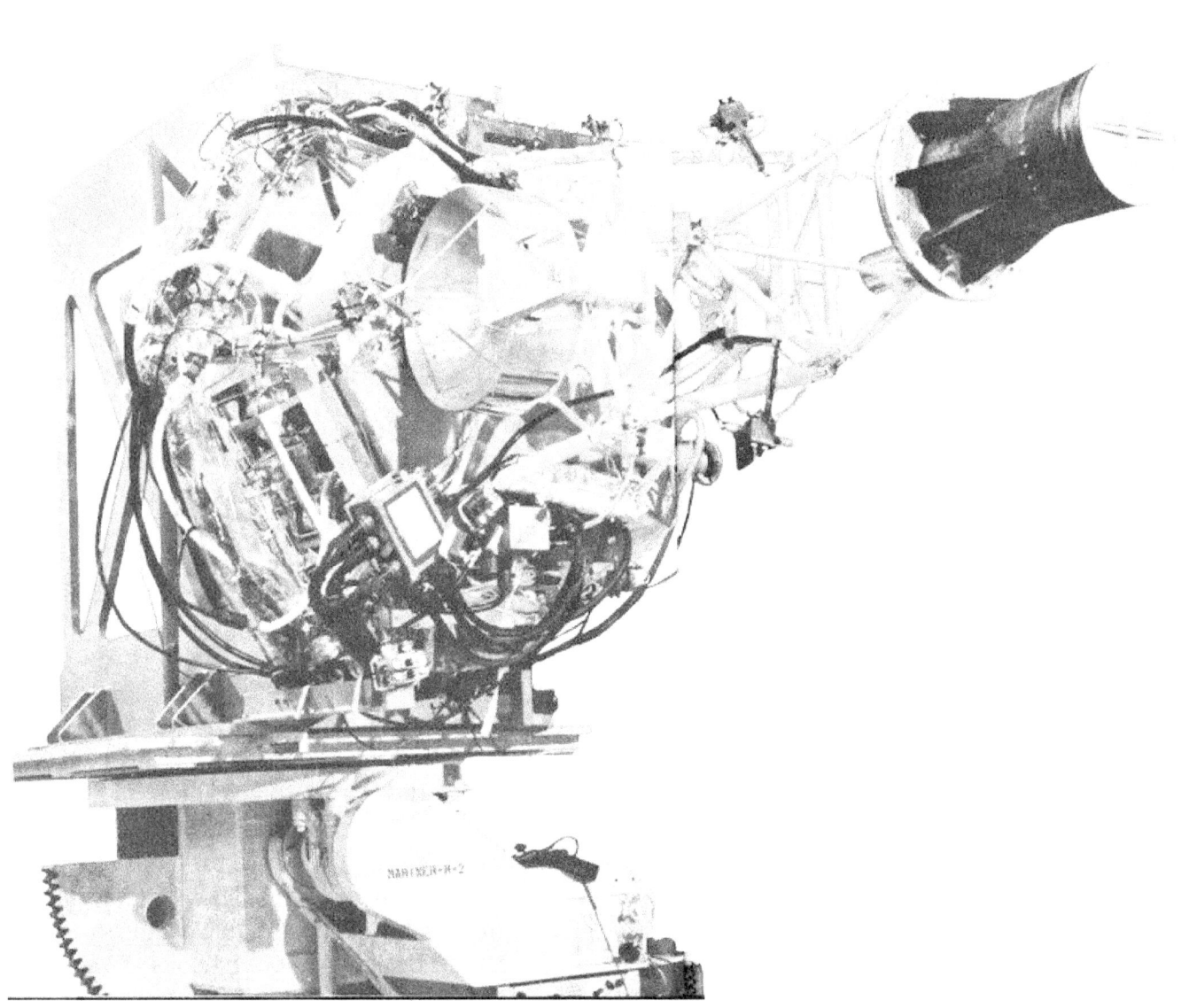

MARINER
MISSION TO VENUS

This book describes one phase of the U. S. civilian space program—the journey of the Mariner spacecraft to the vicinity of Venus and beyond. It reports upon the measurements taken during the "flyby" on December 14, 1962, when Mariner reached a point 21,598 miles from the planet, and 36,000,000 miles from Earth (communication with the spacecraft was continued up to a distance of approximately 54,000,000 miles from Earth). The Mariner mission was a project of the National Aeronautics and Space Administration, carried out under Contract No. NAS 7-100 by the Jet Propulsion Laboratory, California Institute of Technology.

FOREWORD

For many centuries scientific information about the planets and the vast void that separates them has been collected by astronomers observing from the surface of the Earth. Now, with the flight of Mariner II, we suddenly have in our hands some 90 million bits of experimental data measured in the region between Earth and the planet Venus. Thus, man for the first time has succeeded in sending his instruments far into the depths of space, and indeed, in placing them near another planet. A whole new area of experimental astronomy has been opened up.

This book is a brief record of the Mariner Project to date and is designed to explain in general terms the preliminary conclusions. Actually, it will be months or years before all of the data from Mariner II have been completely analyzed. The most important data were the measurements made in the vicinity of the planet Venus, but it should also be noted that many weeks of interplanetary environmental measurements have given us new insight into some of the basic physical phenomena of the solar system. The trajectory data have provided new, more accurate measurements of the solar system. The engineering measurements of the performance of the spacecraft will be of inestimable value in the design of future spacecraft. Thus, the Mariner II spacecraft to Venus not only looks at Venus but gives space scientists and engineers information helpful in a wide variety of space ventures.

A project such as Mariner II is first a vast engineering task. Many thousands of man-hours are required to design the complex automatic equipment which must operate perfectly in the harsh environment of space. Every detail of the system must be studied and analyzed. The operations required to carry out the mission must be understood and performed with precision. A successful mission requires every member of the entire project team to do his task perfectly. Whether it be the error of a designer, mechanic, mathematician, technician, operator, or test engineer—a single mistake, or a faulty piece of workmanship, may cause the failure of the mission. Space projects abound with examples of the old saying, "For want of a nail, the shoe was lost ...," and so on, until the kingdom is lost. Only when every member of the project team is conscious of his responsibility will space projects consistently succeed.

The Mariner II Project started with the Lunar and Planetary Projects Office of the Office of Space Sciences at NASA in Washington. Jet Propulsion Laboratory, California Institute of Technology, personnel provided the main body of the team effort. They were heavily supported by industrial contractors building many of the subassemblies of the spacecraft, by scientists planning and designing the scientific experiments, and by the Air Force which supplied the launching rockets. Several thousand men and women had some direct part in the Mariner Project. It would be impossible to list all of those who made some special contribution, but each and every member of the project performed his job accurately, on time, and to the highest standards.

Mariner II is only a prelude to NASA's program of unmanned missions to the planets. Missions to Mars as well as Venus will be carried out. Spacecraft will not only fly by the planets as did Mariner II, but capsules will be landed, and spacecraft will be put into orbit about the planets. The next mission in the Mariner series will be a flyby of the planet Mars in 1965.

By the end of the decade, where will we be exploring, what will new Mariners have found? Will there be life on Mars, or on any other planet of the solar system? What causes the red spot on Jupiter? What is at the heart of a comet? These and many other questions await answers obtained by our future spacecraft. Mariner II is just a beginning.

W. H. Pickering

Director

Jet Propulsion Laboratory

California Institute of Technology

April, 1963

CONTENTS

ACKNOWLEDGMENTS

Researching the material, gathering and comparing data, preparation of review drafts and attending to the hundreds of details required to produce a document on the results of such a program as the Mariner mission to Venus is a tremendous task. Special acknowledgment is made to Mr. Harold J. Wheelock who, on an extremely short time scale, carried the major portion of this work to completion.

Although the prime sources for the information were the Planetary Program office and the Technical Divisions of the Jet Propulsion Laboratory, other organizations were extremely helpful in providing necessary data, notably the George C. Marshall Space Flight Center, the Lockheed Missiles and Space Company, the Astronautics Division of the General Dynamics Corporation, and, of course, the many elements of the National Aeronautics and Space Administration.

JPL technical information staff members who assisted Mr. Wheelock in production of the manuscript and its illustrations were Mr. James H. Wilson, Mr. Arthur D. Beeman and Mr. Albert E. Tyler. JPL is also grateful to Mr. Chester H. Johnson for his help and suggestions in preparing the final manuscript.

CHAPTER 1

VENUS

Halfway between Los Angeles and Las Vegas, the California country climbs southward out of the sunken basin of Death Valley onto the 3500-foot-high floor of the Mojave desert.

On this immense plateau in an area near Goldstone Dry Lake, about 45 miles north of the town of Barstow, a group of 85-foot antennas forms the nucleus of the United States' world-wide, deep-space tracking network.

Here, on the morning of December 14, 1962, several men were gathered in the control building beneath one of the antennas, listening intently to the static coming from a loudspeaker. They were surrounded by the exotic equipment of the space age. Through the window loomed the gleaming metal framework of an antenna.

Suddenly a voice boomed from the loudspeaker: "The numbers are changing. We're getting data!"

The men broke into a cheer, followed by an expectant silence.

Again the voice came from the speaker: "The spacecraft's crossing the terminator ... it's still scanning."

At that moment, some 36 million miles from the Earth, the National Aeronautics and Space Administration's Mariner[1] spacecraft was passing within 21,600 miles of the planet Venus and was radioing back information to the Goldstone Station—the first scientific data ever received by man from the near-vicinity of another planet.

At the same time, in Washington, D.C., a press conference was in progress. Mr. James E. Webb, Administrator of the National Aeronautics and Space Administration, and Dr. William H. Pickering, Director of the Jet Propulsion Laboratory, stood before a bank of microphones. In a few moments, Dr. Pickering said, the audience would hear the sound of Mariner II as it transmitted its findings back to the Earth.

Then, a musical warble, the voice of Mariner II, resounded in the hall and in millions of radios and television sets around the nation. Alluding to the Greek belief that harmonious sounds accompanied the movement of the planets, Dr. Pickering remarked that this, in truth, was the music of the spheres.

Mariner II had been launched from Cape Canaveral, Florida, on August 27, 1962. Its arrival at Venus was the culmination of a 109-day journey through the strange environment of interplanetary space. The project had gone from the drawing board to the launching pad in less than 11 months. Mariner had taxed the resources and the manpower of the Jet

Propulsion Laboratory, California Institute of Technology; the Atlantic Missile Range centering at Cape Canaveral; theoretical and experimental laboratories at several universities and NASA centers; numerous elements of the aerospace industry; and, of course, NASA management itself.

To the considerable body of engineers scattered around the world from Pasadena to Goldstone to South Africa to Australia, the warble of Mariner was something more than "the music of the spheres." Intercept with Venus was the climax of 109 days of hope and anxiety.

To the world at large, this warbling tone was a signal that the United States had moved ahead—reached out to the planets. Mariner was exploring the future, seeking answers to some of the unsolved questions about the solar system.

THE DOUBLE STAR OF THE ANCIENT WORLD

Venus, the glittering beacon of our solar system, has intrigued man for at least 4,000 years. The Babylonians first mentioned the brilliant planet on clay tablets as early as 2,000 years before Christ. The Egyptians, the Greeks, and the Chinese had thought of Venus as two stars because it was visible first in the morning and then in the evening sky. The Greeks had called the morning star Phosphorus and the evening star Hesperos. By 500 B.C. Pythagoras, the Greek philosopher, had realized that the two were identical.

Galileo discovered the phases of Venus in 1610. Because of the planet's high reflectivity, Copernicus falsely concluded that Venus was either self-luminous or else transparent to the rays of the Sun.

Venus was tracked across the face of the Sun in 1761, from which event the presence of an atmosphere about the planet was deduced because of the fuzzy edges of the image visible in the telescope. Throughout the eighteenth and nineteenth centuries, Venus continued to excite growing scientific curiosity in Europe and America.

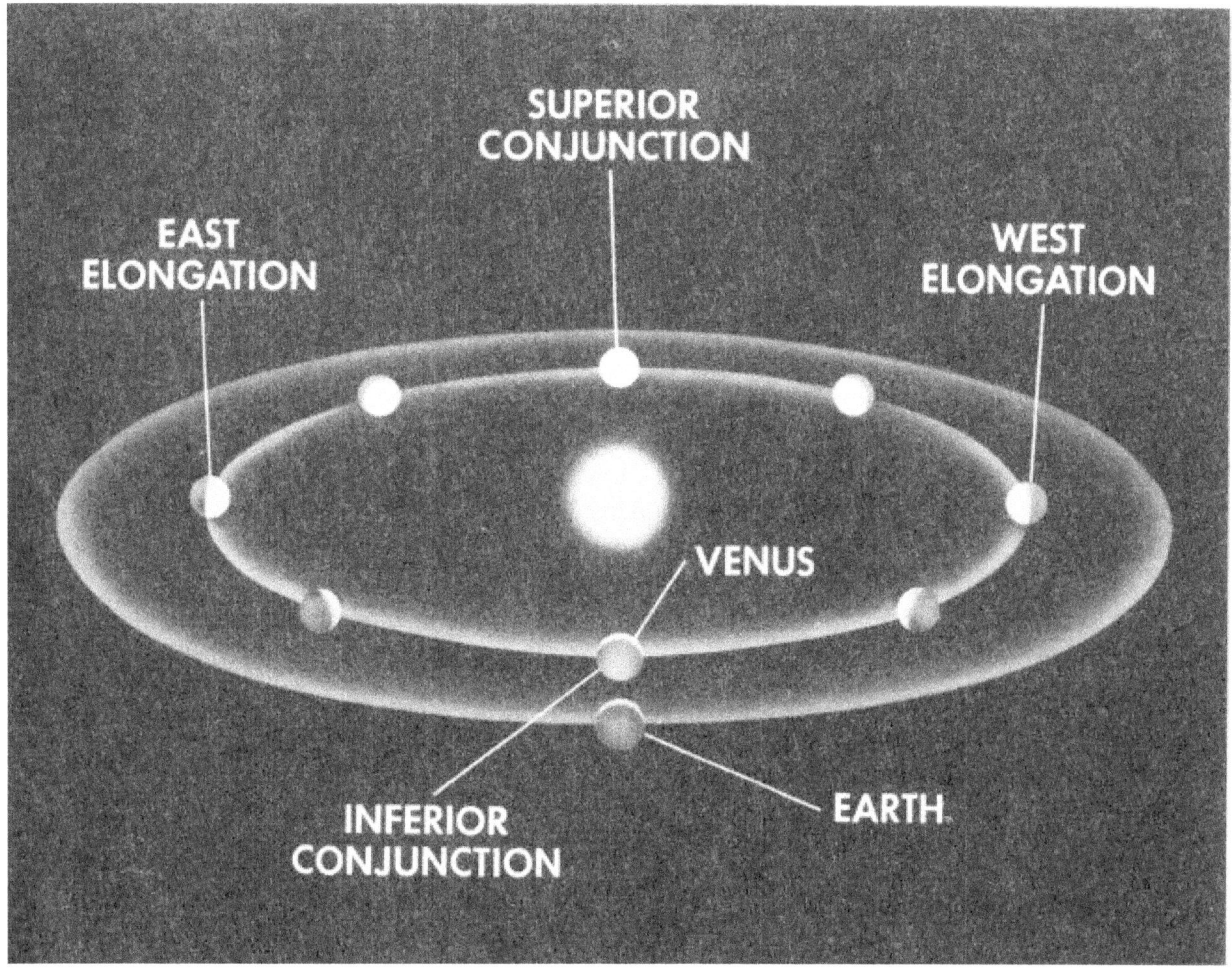

Venus' orbit is almost circular. At inferior conjunction, the planet is between the Earth and the Sun, approximately 26,000,000 miles away; at superior conjunction, Venus is on the other side of the Sun. The elongations are the farthest points to the east and the west of the Earth.

Even the development of giant telescopes and the refinement of spectroscopic and radar astronomy techniques in recent times had yielded few indisputable facts about Venus. Until radar studies, made from Goldstone, California, in 1962, neither the rate nor the angle of axial spin could be determined with any degree of accuracy. The ever-shifting atmosphere continued to shield the Venusian surface from visual observation on Earth, and the nature of its atmosphere became an especially controversial mystery.

THE CONSENSUS PRIOR TO MARINER II

Venus is a virtual twin of the Earth; it approaches our planet closer than any celestial body except the Moon, a few vagrant comets, and other such galactic wanderers. Long fabled in song and legend as the most beautiful object in the sky, Venus has an albedo, or reflectivity factor, of 59% (the Moon has one of 7%). In its brightest or crescent phase, Venus glows like a torch, even casting a distinct shadow—the only body other than the Sun and the Moon yielding such light.

Venus' diameter is approximately 7,700 miles, compared with Earth's 7,900. Also as compared with 1.0 for the Earth, Venus' mean density is 0.91, the mass 0.81, and the volume 0.92.

The Cytherean orbit (the adjective comes from Cytherea, one of the ancient Greek names for Aphrodite—or in Roman times, Venus—the goddess of love) is almost a perfect circle, with an eccentricity (or out-of-roundness) of only 0.0068, lowest of all the planets. Venus rides this orbital path at a mean distance from the Sun of 67.2 million miles (Earth is 93 million miles), and at a mean orbital speed of 78,300 miles per hour, as compared with Earth's 66,600 miles per hour.

It also has a shorter sidereal period (revolution around the Sun or year): 224 Earth days, 16 hours, 48 minutes. Estimates of the Venus rotational period, or the length of the Venus day, have ranged from approximately 23 Earth hours to just over 224 Earth days. The latter rotation rate would be almost equivalent to the Venusian year and, in such case, the planet would always have the same face to the Sun.

Venus approaches within 26 million miles of the Earth at inferior conjunction, and is as far away as 160 million miles at superior conjunction, when it is on the opposite side of the Sun.

The escape velocity (that velocity required to free an object from the gravitational pull of a planet) on Venus is 6.3 miles per second, compared with Earth's escape velocity of 7 miles per second. The gravity of the Earth is sufficient to trap an oxygen-bearing atmosphere near the terrestrial surface. Because the escape velocity of Venus is about the same as that of Earth, men have long believed (or hoped) that the Cytherean world might hold a similar atmosphere and thus be favorable to the existence of living organisms as we know them on the Earth. From this speculation, numerous theories have evolved.

THE CYTHEREAN RIDDLE: LIVING WORLD OR INCINERATED PLANET

Before Mariner II, Venus probably caused more controversy than any other planet in our solar system except Mars. Observers have visualized Venus as anything from a steaming abode of Mesozoic-like creatures such as were found on the Earth millions of years ago, to a dead, noxious, and sunless world constantly ravaged by winds of incredible force.

Conjectures about the Venusian atmosphere have been inescapably tied to theories about the Venusian topography. Because the clouds forming the Venusian atmosphere, as viewed from the Earth through the strongest telescopes, are almost featureless, this relationship between atmosphere and topography has posed many problems.

Impermanent light spots and certain dusky areas were believed by some observers to be associated with Venusian oceans. One scientist believed he identified a mountain peak which he calculated as rising more than 27 miles above the general level of the planet.

Another feature of the Venusian topography is the lack of (detectable) polar flattening. The Earth does have such a flattening at the poles and it was reasoned that, because Venus did not, its rate of rotation must be much slower than that of the Earth, perhaps as little as only once during a Venusian year, thus keeping one face perpetually toward the Sun.

Another school of thought speculated that Venus was covered entirely by vast oceans; other observers concluded that these great bodies of water have long since evaporated and that the winds, through the Cytherean ages, have scooped up the remaining chloride salts and blasted them into the Venusian skies, thus forming the clouds.

Related to the topographic speculations were equally tenuous theories about its atmosphere. It was reasoned that if the oceans of Venus still exist, then the Venusian clouds may be composed of water droplets; if Venus were covered by water, it was suggested that it might be inhabited by Venusian equivalents of Earth's Cambrian period of 500 million years ago, and the same steamy atmosphere could be a possibility.

Other theories respecting the nature of the Venusian atmosphere, depending on how their authors viewed the Venusian terrain, included clouds of hydrocarbons (perhaps droplets of oil), or vapors of formaldehyde and water. Finally, the seemingly high temperature of the planet's surface, as measured by Earth-bound instruments, was credited by some to the false indications that could be given by a Cytherean ionosphere heavily charged with free electrons.

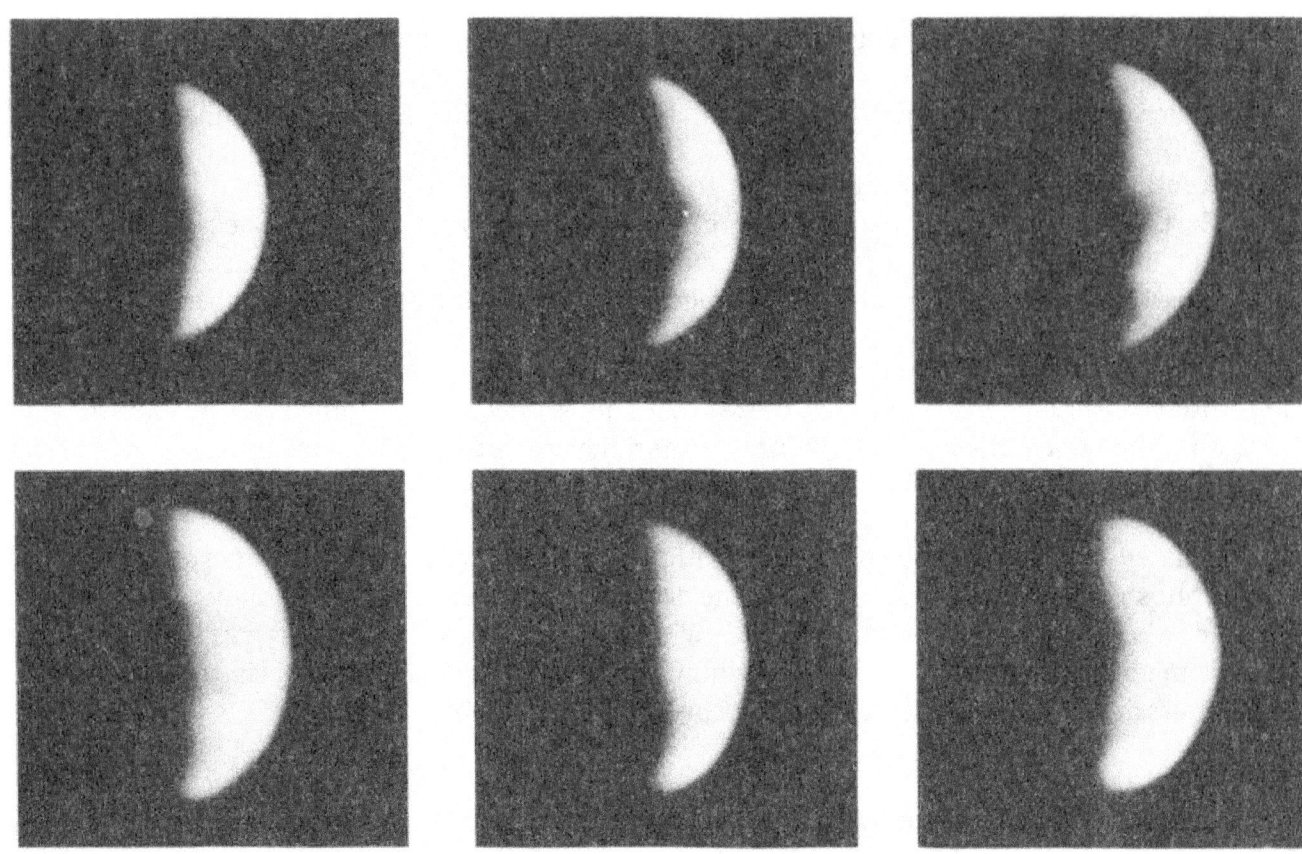

As seen from Earth, Venus is brightest at its crescent phases as shown in these six photographs made by the 100-inch telescope at Mt. Wilson, California.

However, the consensus of pre-Mariner scientific thinking seemed generally to indicate no detectable free oxygen in the atmosphere; this fact inveighed against the probability of surface vegetation, because Earth-bound vegetation, at least, uses carbon dioxide and gives off oxygen into the atmosphere. On the other hand, a preponderance of carbon dioxide in the Venusian atmosphere was measured which would create a greenhouse effect. The heat of the Sun would be trapped near the surface of the planet, raising the temperature to as high as 615 degrees F. If the topography were in truth relatively flat and the rate of rotation slow, the heating effect might produce winds of 400 miles per hour or more, and sand and dust storms beyond Earthly experience. And so the controversy continued.

But at 1:53.13.9 a.m., EST, on August 27, 1962, the theories of the past few centuries were being challenged. At that moment, the night along the east Florida coast was shattered by the roar of rocket engines and the flash of incandescent exhaust streams. The United States was launching Mariner II, the first spacecraft that would successfully penetrate interplanetary space and probe some of the age-old mysteries of our neighbor planet.

CHAPTER 2
PREPARING FOR SPACE

In the summer of 1961, the United States was pushing hard to strengthen its position in the exploration of space and the near planets. The National Aeronautics and Space Administration was planning two projects, both to be launched by an Atlas booster and a Centaur high-energy second stage capable of much better performance than that available from earlier vehicles.

The Mariner program had two goals: Mariner A was ticketed for Venus and Mariner B was scheduled to go to Mars. Caltech's Jet Propulsion Laboratory had management responsibility under NASA for both projects. These spacecraft were both to be in the 1,000- to 1,250-pound class. Launch opportunities for the two planets were to be best during the 1962-1964 period and the new second-stage booster known as Centaur was expected to be ready for these operations.

But trouble was developing for NASA's planners. By August, 1961, it had become apparent that the Centaur would not be flying in time to take advantage of the 1962 third-quarter firing period, when Venus would approach inferior conjunction with the Earth. JPL studied the problem and advised NASA that a proposed lightweight, hybrid spacecraft combining certain design features of Ranger III (a lunar spacecraft) and Mariner A could be launched to Venus in 1962 aboard a lower-powered Atlas-Agena B launch vehicle.

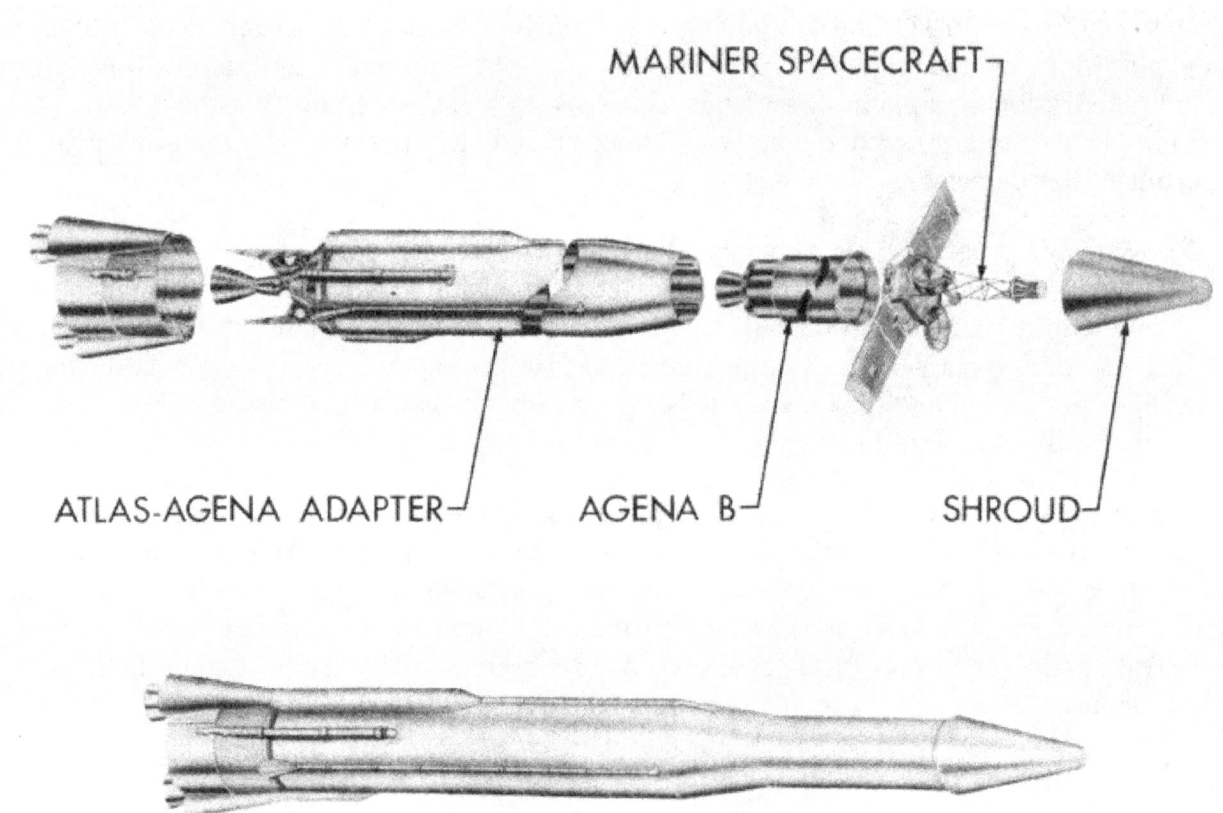

MARINER SPACECRAFT

ATLAS-AGENA ADAPTER

AGENA B

SHROUD

The Mariner II spacecraft was launched by an Atlas first-stage booster vehicle and an Agena B second stage with restart capability.

ATLAS-AGENA ADAPTER

AGENA B

MARINER SPACECRAFT

SHROUD

The proposed spacecraft would be called Mariner R and was to weigh about 460 pounds and carry 25 pounds of scientific instruments (later increased to 40 pounds). The restart capability of Agena was to be used in a 98-statute-mile parking orbit. (The orbit was later raised to 115 statute miles and the spacecraft weight was reduced to about 447 pounds.)

Two spacecraft would be launched one after the other from the same pad within a maximum launch period extending over 56 days from July to September, 1962. The minimum launch separation between the two spacecraft would be 21 days.

As a result of the JPL recommendations, NASA cancelled Mariner A in September, 1961, and assigned JPL to manage a Mariner R Project to fly two spacecraft (Mariner I and II) to the vicinity of Venus in 1962. Scientific measurements were to be made in interplanetary space and in the immediate environs of the planet, which would also be surveyed in an attempt to determine the characteristics of its atmosphere and surface. Scientific and engineering data would also be transmitted from the spacecraft to the Earth while it was in transit and during the encounter with Venus.

Scientists and engineers were now faced with an arduous task. Within an 11-month period, on a schedule that could tolerate no delays, two spacecraft had to be designed, developed, assembled, tested, and launched. In order to meet the schedule, tested flight assemblies and instruments would have to be in the Pasadena assembly facility by mid-January, 1962, just four months after the start of the project. Probably no other major space project of similar scope had ever been planned on such a demanding schedule.

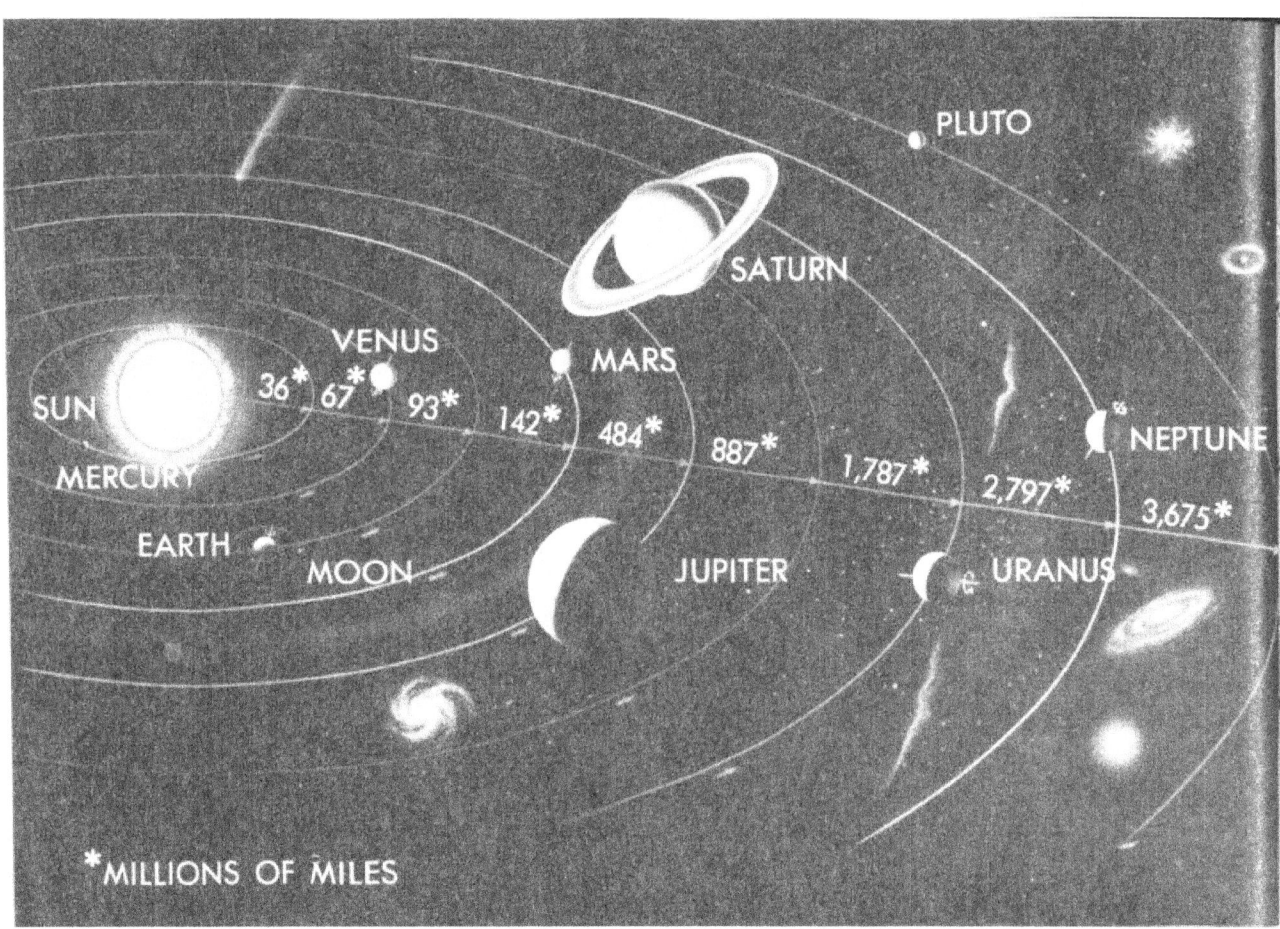

Mariner II travelled across 180 million miles of space within our solar system as it spanned the gap between Earth and Venus (shown here as the third and second planets, respectively, from the Sun).

With the shipment of equipment to Atlantic Missile Range (AMR) scheduled for 9½ months after inception of the project, management and design teams went all-out on a true "crash" effort. Quick decisions had to be made, a workable design had to be agreed upon very early, and, once established, the major schedule objectives could not be changed. Certain design modifications and manufacturing changes in the Atlas-Agena launch vehicle were also necessary.

Wherever possible, Ranger design technology had to be used in the new spacecraft and adapted to the requirements of a planetary probe. Other necessary tasks included trajectory calculation; arrangements for launch, space flight, and tracking operations; and coordination of AMR Range support.

Following NASA's September, 1961 decision to go ahead with the Mariner R Project, JPL's Director, Dr. William H. Pickering, called on his seasoned team of scientists and engineers. Under Robert J. Parks, Planetary Program Director, Jack N. James was appointed as Project Manager for Mariner R, assisted by W. A. Collier. Dan Schneiderman was appointed Spacecraft System Manager, and Dr. Eberhardt Rechtin headed the space tracking program, with supervision of the Deep Space Instrumentation Facility (DSIF) operations under Dr. Nicholas Renzetti. The Mariner space flight operations were directed by Marshall S. Johnson.

A PROBLEM IN CELESTIAL DYNAMICS

In order to send Mariner close enough to Venus for its instruments to gather significant data, scientists had to solve aiming and guidance problems of unprecedented magnitude and complexity.

The 447-pound spacecraft had to be catapulted from a launching platform moving around the Sun at 66,600 miles per hour, and aimed so precisely that it would intercept a planet moving 78,300 miles per hour (or 11,700 miles per hour faster than the Earth) at a point in space and time some 180.2 million miles away and 109 days later, with only one chance to correct the trajectory by a planned midcourse maneuver.

And the interception had to be so accurate that the spacecraft would pass Venus within 8,000 to 40,000 miles. The chances of impacting the planet could not exceed 1 in 1,000 because Mariner was not sterilized and might contaminate Venus. Also, much more data could be gathered on a near-miss flight path than on impact. Furthermore, at encounter (in the target area) the spacecraft had to be so positioned that it could communicate with Earth, see the Sun with its solar panels, and scan Venus at the proper angles.

Along the way, Mariner had to be able to orient itself so that its solar panels were facing or "locked onto" the Sun in order to generate its own power; acquire and maintain antenna orientation to the Earth; correct its attitude constantly to hold Earth and Sun lock; receive, store, and execute commands to alter its course for a closer approach to Venus; and communicate its findings to Earth with only 3 watts of radiated power and over distances never before spanned.

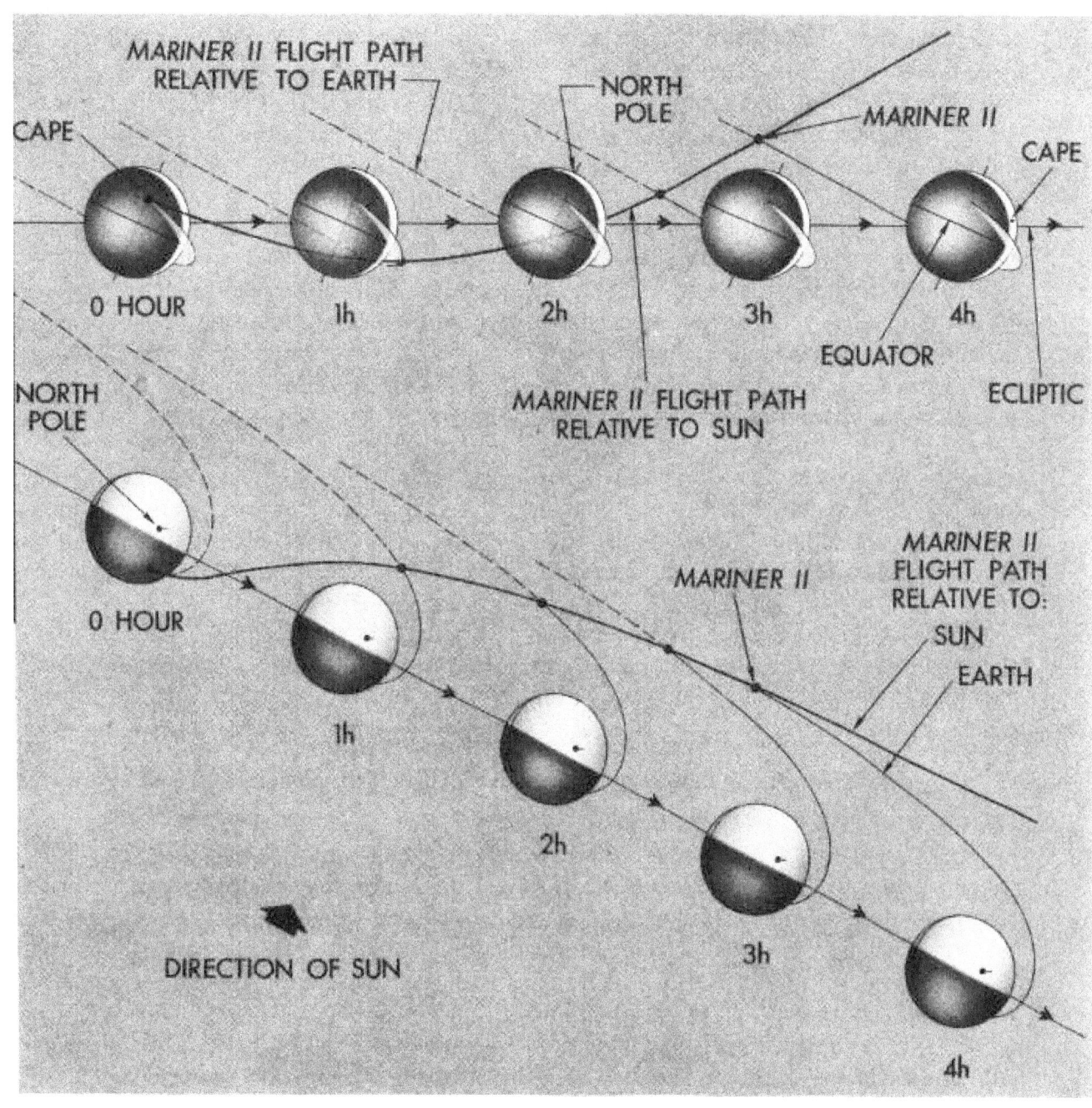

Mariner II was launched in a direction opposite to the orbital travel of the Earth. The Sun's gravity then pulled it in toward the planet Venus.

Early in the program it had been decided that two spacecraft would be launched toward Venus. Only 56 days were available for both launchings and the planet would not be close enough again for 19 months—the period between inferior conjunctions or the planet's closest approach to the Earth. On any one of these days, a maximum of 2 hours could be used for getting the vehicles off the launch pad. In addition, the Mariners would have to leave the Earth in a direction opposite to that of the Earth's direction of orbital revolution around the Sun. This flight path was necessary so the spacecraft could then fall in toward the Sun and intercept Venus, catching and passing the Earth along the way, about 65 days and 11.5 million miles out.

This feat of celestial navigation had to be performed while passing through the hostile environment of interplanetary space, where the probe might be subjected to solar winds (charged particles) travelling at velocities up to 500 miles per second; intense bombardment from cosmic radiation, charged protons, and alpha particles moving perhaps 1.5 million miles per hour; radiated heat that might raise the spacecraft temperatures to unknown values; and the unknown dangers from cosmic dust, meteorites, and other miscellaneous space debris.

In flight, each spacecraft would have to perform more than 90,000 measurements per day, reporting back to the Earth on 52 engineering readings, the changes in interplanetary magnetic fields, the density and distribution of charged particles and cosmic dust, and the intensity and velocity of low-energy protons streaming out from the Sun.

At its closest approach to Venus, the spacecraft instruments would be required to scan the planet during a brief 35-minute encounter, to gather data that would enable Earth scientists to determine the temperature and structure of the atmosphere and the surface, and to process and transmit that data back to the Earth.

THE ORGANIZATION

Flying Mariner to Venus was a team effort made possible through the combined resources of several United States governmental organizations and their contractors, science, and industry. The success of the Mariner Project resulted primarily from the over-all direction and management of the National Aeronautics and Space Administration and the Jet Propulsion Laboratory, and the production and launch capabilities of the vehicle builders and the Air Force. Several organizations bore the major responsibility: NASA Headquarters, JPL, NASA's Marshall Space Flight Center and Launch Operations Center, Astronautics Division of General Dynamics, and Lockheed Missiles and Space Company.

NASA: FOR SCIENCE

The National Aeronautics and Space Administration was an outgrowth of the participation of the United States in the International Geophysical Year program and of the nation's space effort, revitalized following Russia's successful orbiting of Sputnik I in 1957.

Final NACA meeting, August 21, 1958.

Model of X-1 research plane.

Headquarters of National Aeronautics and Space Administration, Washington, D.C.

JPL developed first JATO units in 1941.

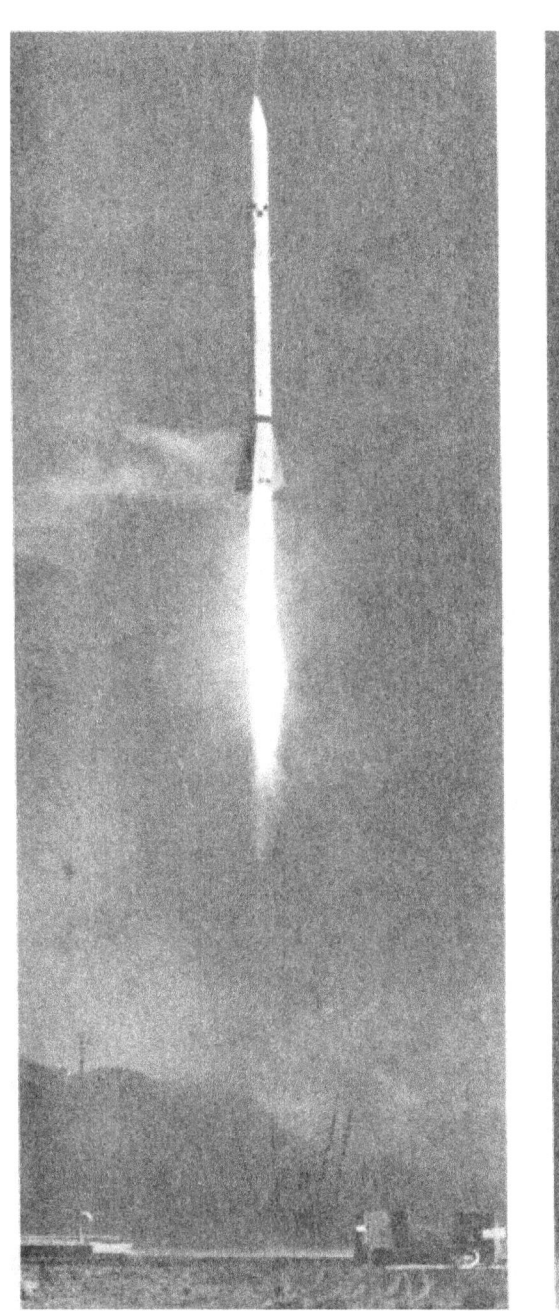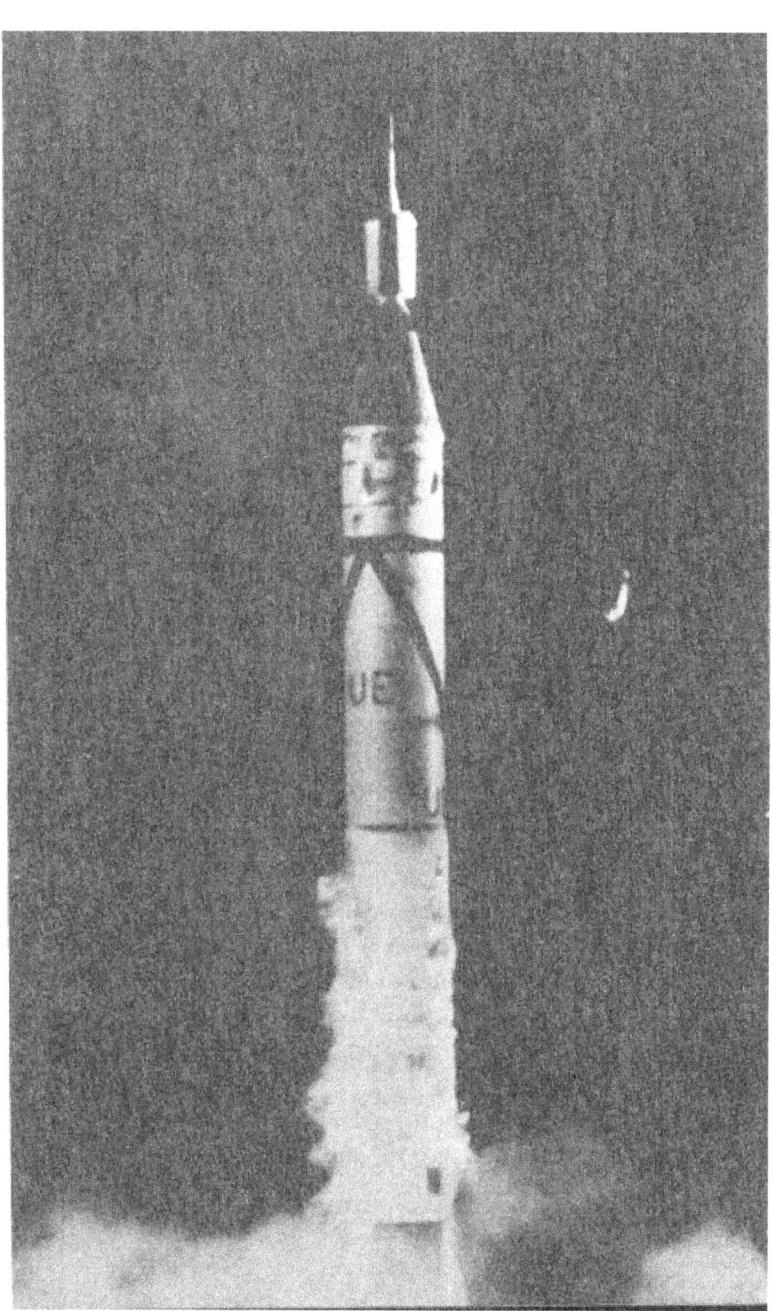

Other Laboratory Projects were the Corporal missile (left) and Explorer I (right), the first U.S. satellite.

Under the terms of the law which created NASA, it is a Federal Agency dedicated to carrying out "activities in space ... devoted to peaceful purposes for the benefit of all mankind." NASA is charged to preserve the role of our nation as a leader in the aeronautical and space sciences and technology and to utilize effectively the science and engineering resources of the United States in accomplishing these goals. Activities associated with military operations in space and the development of weapons systems are specifically assigned to the Defense Department.

In November, 1957, before the creation of NASA, President Eisenhower had established a Scientific Advisory Committee to determine the national objectives and requirements in space and to establish the basic framework within which science, industry, and the academic community could best support these objectives.

The Committee submitted a report to the President in March, 1958, recommending creation of a civilian agency to conduct the national space programs. The recommendation, endorsed by the President, was submitted to the Congress on April 2, 1958. The National Aeronautics and Space Act of 1958 was passed and became law in July, 1958.

NASA was officially established on October 1, 1958, and Dr. T. Keith Glennan, President of Case Institute of Technology, was appointed as the first Administrator. The facilities and personnel of the National Advisory Committee for Aeronautics (NACA) were transferred to form the nucleus of the new NASA agency.

NACA had performed important and significant research in aeronautics, wind tunnel technology, and aerodynamics since 1915, including a series of experimental rocket research aircraft that culminated in the X-15. It was natural that it be expanded to include space operations.

Among the NACA Centers transferred to NASA were the Langley Research Center at Hampton, Virginia; Lewis Research Center, Cleveland, Ohio; Ames Research Center, Moffett Field, California; Flight Research Center, Edwards, California; and the rocket launch facility at Wallops Island, Virginia.

Those personnel of the Naval Research Laboratory who had been working on Project Vanguard were also transferred to NASA, as was the project. These personnel are now part of the new Goddard Space Flight Center at Greenbelt, Maryland.

The October, 1958, transfers also included a number of the space projects of the Advanced Research Projects Agency of the Defense Department. In a December, 1958, Executive Order, the President assigned the former Army facilities of the Jet Propulsion Laboratory at Pasadena, California, to NASA. At the same time, the group working under Dr. Wernher von Braun at the Army Ballistic Missile Agency (commanded by Major General John B. Medaris) was made responsive to NASA requirements.

On July 1, 1960, the George C. Marshall Space Flight Center (MSFC) was organized at Huntsville under von Braun's direction. The former Development Operations Division of ABMA formed the nucleus of the new Center. The MSFC mission was to procure and to supervise the adaptation of launch vehicles for NASA space missions, including Atlas, Thor, and Agena. Marshall is directly responsible for the design and development of advanced, high-thrust booster vehicles such as the Saturn C-1 and C-5 and the Nova.

An agency to conduct NASA affairs at Cape Canaveral was formed within MSFC on July 1, 1960. Known then as the Launch Operations Directorate (LOD), it was directed by Dr. Kurt H. Debus. LOD became independent of Marshall in March, 1962, when it was redesignated the Launch Operations Center (LOC), reporting directly to the Office of Manned Space Flight. This separation resulted largely because the activities at AMR were becoming more operational in character and less oriented toward research and development.

LOC handles such functions for NASA as the scheduling of launch dates and liaison with the Atlantic Missile Range for support activities. The Center will have the responsibility in the field for assembly, checkout, and launch of the Saturn and Nova boosters.

Following the election of President Kennedy in 1961, James E. Webb replaced Dr. Glennan as Administrator of NASA. Shortly after, a new national goal was announced—placing a man on the Moon and returning him safely to the Earth in this decade. Meanwhile, JPL had been assigned responsibility for unmanned exploration of the Moon, the planets, and interplanetary space, and thus was charged with supporting the NASA manned flight program through these activities.

In less than five years, NASA grew to include eight flight and research centers and about 21,000 technical and management personnel. Within NASA, Dr. Abe Silverstein's Office of Space Flight Programs was responsible for the Mariner R Project which was directly assigned to Ed Cortright, Director of Lunar & Planetary Programs, and Fred Kochendorfer, who is NASA's Program Chief for Mariner. A subsequent reorganization placed responsibility under Dr. Homer Newell's Office of Space Sciences, and Oran Nicks became Director of Lunar & Planetary Programs.

JPL: JATO TO MARINER

The Jet Propulsion Laboratory, staffed and operated for NASA by California Institute of Technology, had long been active in research and development in the fields of missiles, rockets, and the space-associated sciences. The first government-sponsored rocket research group in the United States, JPL had originated on the Caltech campus in 1939, an outgrowth of the Guggenheim Aeronautical Laboratories, then headed by celebrated aerodynamicist Dr. Theodore von Karman.

Von Karman and his associates moved their operation to a remote spot at the foot of the San Gabriel mountains and, working from this base, in 1941 the pioneering group developed the first successful jet-assisted aircraft takeoff (JATO) units for the Army Air Force. The Laboratory began a long association with the Army Ordnance Corps in 1944, when the Private A test rocket was developed. In retrospect, it is now recognized that the

Private A was the first U. S. surface-to-surface, solid-propellant rocket. Its range was 10 miles!

JPL's WAC Corporal rocket set a U. S. high-altitude record of 43.5 miles in 1945. Mounted on a German V-2 as the Bumper-WAC, it achieved an altitude record of 250 miles in 1947. More important, this event was the first successful in-flight separation of a two-stage rocket—the feasibility of space exploration had been proved.

After the end of World War II, JPL research set the stage for high-energy solid-propellant rockets. For the first time the solid propellants, which contained both fuel and oxidizers, were cast in thin-walled cases. Techniques were then developed for bonding the propellants to the case, and burning radially outward from the central axis was achieved. Attention was then turned to increasing the energy of the propellants.

By 1947, the Corporal E, a new liquid-propellant research rocket, was being fired. JPL was asked to convert it into a tactical weapon in 1949. The Corporal E then became the first liquid-propellant surface-to-surface guided missile developed by the United States or the Western bloc of nations.

Because of the need for higher mobility and increased firing rate, JPL later designed and developed the solid-propellant Sergeant—the nation's first "second-generation" weapon system. This inertially guided missile was immune to electronic countermeasures by an enemy.

Meanwhile, JPL scientists had pioneered in the development of electronic telemetering techniques, which permit an accurate monitoring of system performance while missiles are in flight. By 1944, Dr. William H. Pickering, a New Zealand born and Caltech-trained physicist who had worked with Dr. Robert Millikan in cosmic ray research, had been placed in charge of the telemetering effort at JPL. Pickering became Director of the Laboratory in 1954.

Following the launching of Sputnik I, the Army-JPL team which had worked on the Jupiter C missile to test nose cones, was assigned the responsibility for putting the first United States satellite into orbit as soon as possible. In just 83 days, a modified Jupiter C launch vehicle was prepared, an instrumented payload was assembled, a network of space communications stations was established, and Explorer I was orbited on January 31, 1958. Explorer was an instrumented assembly developed by JPL and the State University of Iowa. It discovered the inner Van Allen radiation belt.

Subsequently, JPL worked with the Army on other projects to explore space and to orbit satellites. Among these were Pioneer III, which located the outer Van Allen Belt, and Pioneer IV, the first U. S. space probe to reach Earth-escape velocity and to perform a lunar fly-by mission.

The launch vehicle for Mariner was an Atlas D booster with an Agena B second stage. Historically, Atlas can be traced to October, 1954, when the former Convair Corporation (later acquired by General Dynamics) was invited to submit proposals for research and development of four missile systems, including a 5,000-mile intercontinental weapon.

In January, 1946, Convair assigned K. J. Bossart to begin a study of two proposed types of 5,000-mile missiles: one jet powered at subsonic speeds, with wings for aerodynamic control; the other a supersonic, ballistic (wingless and bullet-like), rocket-powered missile capable of operating outside the Earth's atmosphere.

Atlas missiles in assembly facility at General Dynamics/Astronautics plant.

This was the beginning of Project MX-774, lineal ancestor of Atlas. After captive testing at San Diego in 1947, three of the experimental missiles were test-launched at White Sands Proving Ground in New Mexico. The first flight failed at 6,200 feet after a premature engine burnout.

In 1947, the Air Force shelved the MX-774 project. However, this brief program had proved the feasibility of three concepts later used in Atlas: swiveling engines for directional control; lightweight, pressurized airframe structures; and separable nose cones.

The Korean War stimulated the ICBM concept and, in 1951, a new MX-1593 contract was awarded to Convair to study ballistic and glide rockets. By September, 1951, Convair was proposing a ballistic missile that would incorporate some of the features of the MX-774 design. A plan for an accelerated program was presented to the Air Force in 1953. After a year of study, a full go-ahead for the project, now called Atlas, was given in January, 1955.

The unit handling the Atlas program was set up as Convair Astronautics, with J. R. Dempsey as president, on March 1, 1957.

The first Atlas test flight, in June of 1957, ended in destruction of the missile when it went out of control. Following another abortive attempt, the first fully successful flight of an Atlas missile was made from Cape Canaveral on December 17, 1957.

The Atlas program was in full swing by 1958, when 14 test missions were flown. The entire missile was orbited in December, 1958, as Project Score. It carried the voice of President Eisenhower as a Christmas message to the world. The Atlas missile system was accepted for field operations by the Air Force in 1958.

Also in 1958, an Atlas achieved a new distance record, flying more than 9,000 miles down the Atlantic Missile Range, where it landed in the Indian Ocean, off the South African coast.

Atlas has been modified for use by NASA as a space vehicle booster. Known as the Atlas D, it has launched lunar probes, communications and scientific Earth satellites, and manned space vehicles.

LOCKHEED: AGENA B

The Lockheed Agena B second-stage vehicle was mounted on top of the Atlas booster in the launch of the Mariner spacecraft. The U. S. Air Force had first asked Lockheed Missiles and Space Division, headed by L. E. Root, to work on an advanced orbital vehicle for both military and scientific applications in 1956. On October 29 of that year, Lockheed was appointed prime weapon system contractor on the new Agena Project, under the Air Force Ballistic Missile Division. In order to speed the program, the Thor missile was used as the booster stage for the early Agena flights. The Atlas was also utilized in later operations.

In August, 1957, the Air Force recommended that the program be accelerated as much as possible. After Russia orbited Sputnik I in October of 1957, a further speed-up was ordered.

The first of the Agena-Discoverer series was launched into orbit on February 28, 1959, with the Thor missile as the booster. The first restart in orbit occurred on February 18, 1961, when the new Agena B configuration was used to put Discoverer XXI into orbit. All of the NASA missions using Agena, beginning with Ranger I in August, 1961, have been flown with the B model.

Agena holds several orbiting records for U. S. vehicles. The first water recovery followed the 17 orbits of Discoverer XIII on August 11, 1960. The first air recovery of a capsule from orbit occurred with Discoverer XIV on August 18, 1960. In all, a total of 11 capsules were recovered from orbit, 7 in the air, 4 from the sea.

CHAPTER 3
THE SPACECRAFT

In the 11 brief months which JPL had to produce the Mariner spacecraft system, there was no possibility of designing an entirely new spacecraft. JPL's solution to the problem was derived largely from the Laboratory's earlier space exploration vehicles, such as the Vega, the Ranger lunar series, and the cancelled Mariner A.

Wherever possible, components and subsystems designed for these projects were either utilized or redesigned. Where equipment was purchased from industrial contractors, existing hardware was adapted, if practicable. Only a minimum of testing could be performed on newly designed equipment and lengthy evaluation of "breadboard" mock-ups was out of the question.

Ready for launch, the spacecraft measured 5 feet in diameter and 9 feet 11 inches in height. With the solar panels and the directional antenna unfolded in the cruise position, Mariner was 16 feet 6 inches wide and 11 feet 11 inches high.

THE SPACEFRAME

The design engineers were forced to work within the framework of the earlier spacecraft technology because of the time restrictions, but Mariner I and II could weigh only about half as much as the Ranger spacecraft and just over one-third as much as the planned Mariner A.

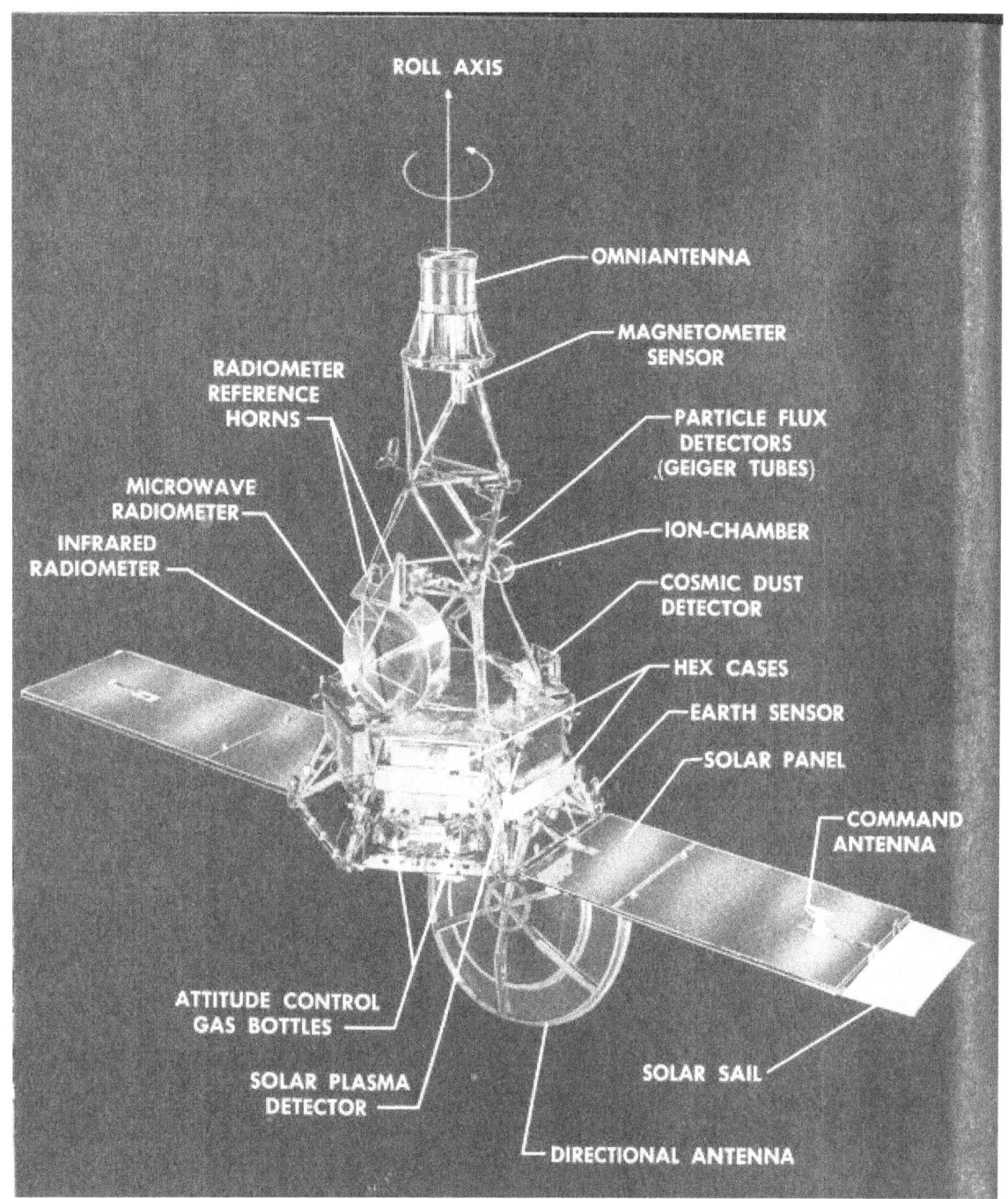

Mariner spacecraft with solar panels, microwave radiometer, and directional antenna extended in flight position. Principal components are shown.

ROLL AXIS

OMNIANTENNA

MAGNETOMETER SENSOR

PARTICLE FLUX DETECTORS (GEIGER TUBES)

RADIOMETER REFERENCE HORNS

MICROWAVE RADIOMETER

INFRARED RADIOMETER

ION-CHAMBER

COSMIC DUST DETECTOR

EARTH SENSOR

SOLAR PANEL

COMMAND ANTENNA

SOLAR SAIL

ATTITUDE CONTROL GAS BOTTLES

SOLAR PLASMA DETECTOR

DIRECTIONAL ANTENNA

The basic structural unit of Mariner was a hexagonal frame made of magnesium and aluminum, to which was attached an aluminum superstructure, a liquid-propelled rocket engine for midcourse trajectory correction, six rectangular chassis mounted one on each face of the hexagonal structure, a high-gain directional antenna, the Sun sensors, and gas jets for control of the spacecraft's attitude.

The tubular, truss-type superstructure extended upward from the base hexagon. It provided support for the solar panels while latched under the shroud during the launch phase, and for the radiometers, the magnetometer, and the nondirectional antenna, which was mounted at the top of the structure. The superstructure was designed to be as light as possible, yet be capable of withstanding the predicted load stresses.

The six magnesium chassis mounted to the base hexagon housed the following equipment: the electronics circuits for the six scientific experiments, the communications system electronics; the data encoder (for processing data before telemetering it to the Earth) and the command electronics; the attitude control, digital computer, and timing sequencer circuits; a power control and battery charger assembly; and the battery assembly.

The allotment of weights for Mariner II forced rigid limitation in the structural design of the spacecraft. As launched, the weights of the major spacecraft subsystems were as follows:

Structure	77 pounds
Solar panels	48 pounds
Electronics	146 pounds
Propulsion	32 pounds
Battery	33 pounds
Scientific experiments	41 pounds
Miscellaneous equipment	70 pounds
Gross weight	447 pounds

THE POWER SYSTEM

Mariner II was self-sufficient in power. It converted energy from sunlight into electrical current through the use of solar panels composed of photoelectric cells which charged a battery installed in one of the six chassis on the hexagonal base. The control, switching, and regulating circuits were housed in another of the chassis cases.

This hexagonal frame, constructed of magnesium and aluminum, is the basic supporting structure around which the Mariner spacecraft is assembled.

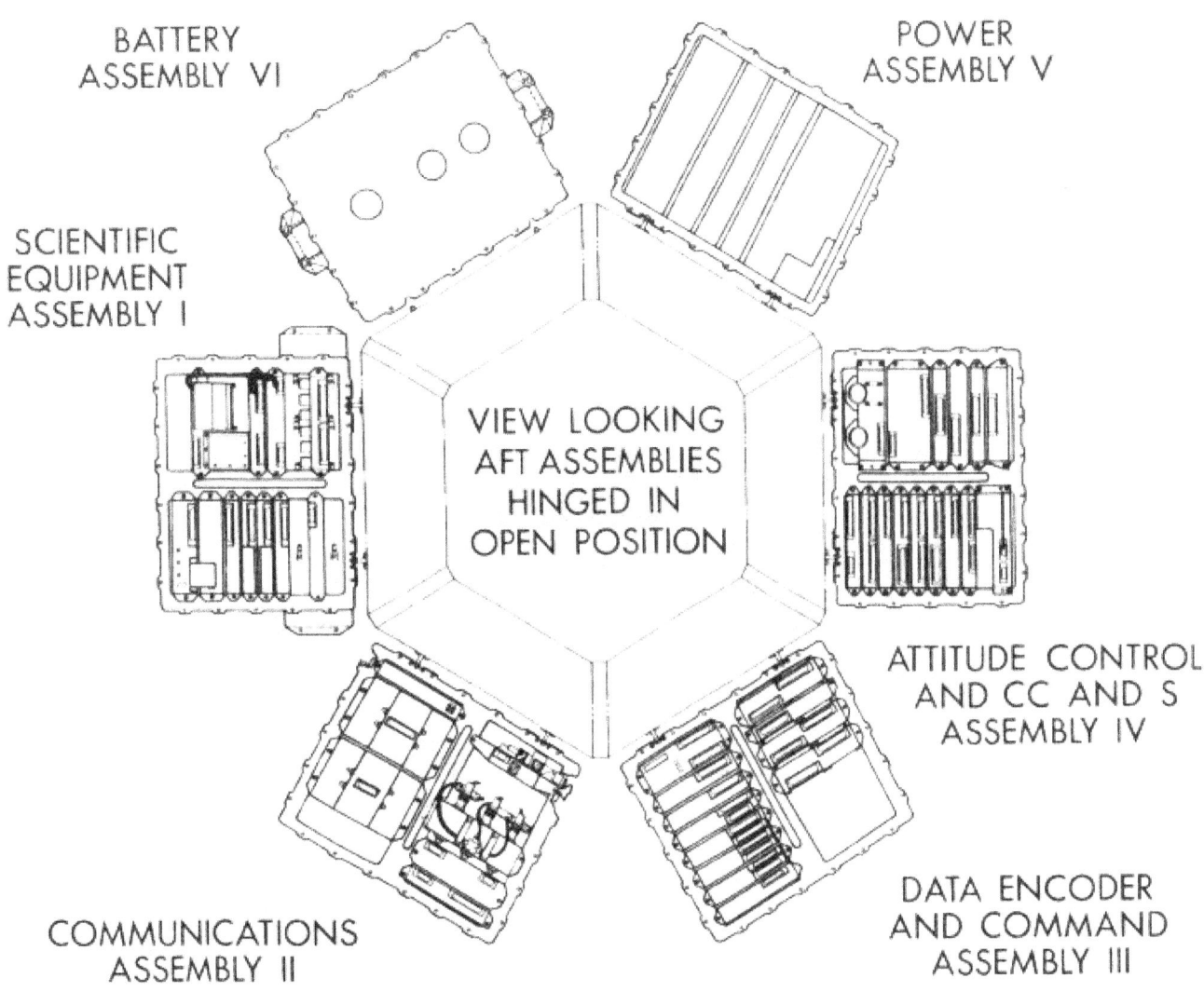

Plan view from top showing six magnesium chassis hinged in open position.

VIEW LOOKING AFT ASSEMBLIES HINGED IN OPEN POSITION

SCIENTIFIC EQUIPMENT ASSEMBLY I

COMMUNICATIONS ASSEMBLY II

DATA ENCODER AND COMMAND ASSEMBLY III

ATTITUDE CONTROL AND CC AND S ASSEMBLY IV

POWER ASSEMBLY V

BATTERY ASSEMBLY VI

The battery operated the spacecraft systems during the period from launch until the solar panels were faced onto the Sun. In addition, the battery supplied power during trajectory maneuvers when the panels were temporarily out of sight of the Sun. It shared the demand for power when the panels were overloaded. The battery furnished power directly for switching various equipment in flight and for certain other heavy loads of brief duration, such as the detonation of explosive devices for releasing the solar panels.

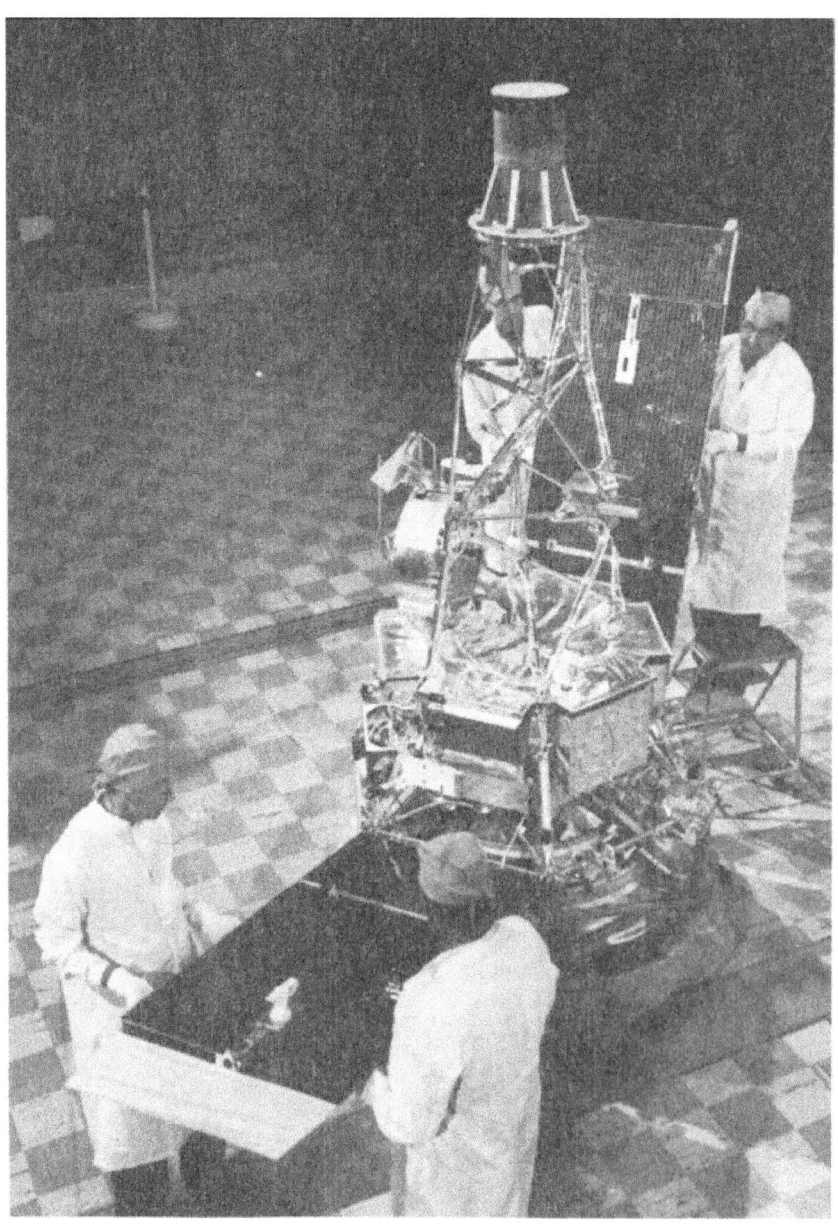

Mariner spacecraft with solar panels in open position. Note extension to left panel to balance solar pressures in flight.

The Mariner battery used sealed silver-zinc cells and had a capacity of 1000 watt-hours. It weighed 33 pounds and was recharged in flight by the solar panels.

The solar panels, as originally designed, were 60 inches long by 30 inches wide and contained approximately 9800 solar cells in a total area of 27 square feet. Each solar cell produced only about 230 one-thousandths of a volt. The entire array was designed to convert the Sun's energy to electrical power in the range between 148 and 222 watts. When a later design change required the extension of one panel in order to add more solar cells, it was necessary to add a blank extension to the other panel in order to balance the solar pressure on the spacecraft.

In order to protect the solar cells from the infrared and ultraviolet radiation of the Sun, which would produce heat but no electrical energy, each cell was shielded from these rays by a glass filter which was nevertheless transparent to the light which the cells converted into power.

The power subsystem electronics circuits were housed in another of the hexagon chassis cases. This equipment was designed to receive and switch power either from the solar panels, the battery, or a combination of the two, to a booster-regulator.

CC&S: THE BRAIN AND THE STOPWATCH

Once the Atlas booster lifted Mariner off the launch pad, the digital Central Computer and Sequencer (CC&S) performed certain computations and provided the basic timing control for those spacecraft subsystems which required a sequenced programming control.

The CC&S was designed to initiate the operations of the spacecraft in three distinct sequences or "modes": (1) the launch mode, from launch through the cruise configuration; (2) the midcourse propulsion mode, when Mariner readjusted its sights on Venus; and (3) the encounter mode, involving commands for data collection in the immediate vicinity of the planet.

The CC&S timed Mariner's actions as it travelled more than 180 million miles in pursuit of Venus. A highly accurate electronic clock (crystal-controlled oscillator) scheduled the operations of the spacecraft subsystems. The oscillator frequency of 307.2 kilocycles was reduced to the 2,400- and 400-cycle-per-second output required for the power subsystem.

The control oscillator also timed the issuance of commands by the CC&S in each of the three operating modes of the spacecraft.

A 1-pulse-per-minute signal was provided for such launch sequence events as the extension of the solar panels 44 minutes after launch, turning on power for the attitude control subsystem one hour after launch, and for certain velocity correction commands during the midcourse maneuver.

The spacecraft used two antennas for communication. The omni-antenna (top) was utilized when the directional antenna (bottom) could not be pointed at the Earth.

This command antenna (on solar panel) was used to receive maneuver commands.

A 1-pulse-per-second signal was generated as a reference during the roll and pitch maneuvers in the midcourse trajectory correction phase. One pulse was generated every 3.3 hours in order to initiate the command to orient the directional antenna on the Earth at 167 hours after launch.

Finally, one pulse every 16.7 hours was used to readjust the Earth-oriented direction of the antenna throughout the flight.

TELECOMMUNICATIONS: RELAYING THE DATA

The telecommunications subsystem enabled Mariner to receive and to decode commands from the Earth, to encode and to transmit information concerning space and Mariner's own functioning, and to provide a means for precise measurement of the spacecraft's velocity and position relative to the Earth. The spacecraft accomplished all these functions using only 3 watts of transmitted power up to a maximum range of 53.9 million miles.

A data encoder unit, with CC&S sequencing, timed the three phases of Mariner's journey: (1) In the launch mode, only engineering data on spacecraft performance were transmitted;

(2) during the cruise mode, information concerning space and Mariner's own functioning was transmitted; and (3) while the spacecraft was in the vicinity of Venus, only scientific information concerning the planet was to be transmitted. (The CC&S failed to start the third mode automatically and it was initiated by radio command from the Earth.) After the encounter with Venus, Mariner was programmed to switch back to the cruise mode for handling both engineering and science data (this sequence was also commanded by Earth radio).

Mariner II used a technique for modulating (superimposing intelligent information) its radio carrier with telemetry data known as phase-shift keying. In this system, the coded signals from the telemetry measurements displace another signal of the same frequency but of a different phase. These displacements in phase are received on the Earth and then translated back into the codes which indicate the voltage, temperature, intensity, or other values measured by the spacecraft telemetry sensors or scientific instruments.

A continually repeating code, almost noise-like both in sound and appearance on an oscilloscope, was used for synchronizing the ground receiver decoder with the spacecraft. This decoder then deciphered the data carried on the information channel.

This technique was called a two-channel, binary-coded, pseudo-noise communication system and it was used to modulate a radio signal for transmission, just as in any other radio system.

Radio command signals transmitted to Mariner were decoded in a command subassembly, processed, and routed to the proper using devices. A transponder was used to receive the commands, send back confirmation of receipt to the Earth, and distribute them to the spacecraft subsystems.

Mariner II used four antennas in its communication system. A cone-like nondirectional (omni) antenna was mounted at the top of the spacecraft superstructure, and was used from injection into the Venus flight trajectory through the midcourse maneuver (the directional antenna could not be used until it had been oriented on the Earth).

A dish-type, high-gain, directional antenna was used at Earth orientation and after the trajectory correction maneuver was completed. It could receive radio signals at greater distances than the nondirectional antenna. The directional antenna was nested beneath the hexagonal frame of the spacecraft while it was in the nose-cone shroud. Following the unfolding of the solar panels, it was swung into operating position, although it was not used until after the spacecraft locked onto the Sun.

The directional antenna was equipped with flexible coaxial cables and a rotary joint. It could move in two directions; one motion was supplied by rolling the spacecraft around its long axis.

In addition, two command antennas, one on either side of one of the solar panels, received radio commands from the Earth for the midcourse maneuver and other functions.

ATTITUDE CONTROL: BALANCING IN SPACE

Mariner II had to maintain a delicate balance in its flight position during the trip to Venus (like a tight-wire walker balancing with a pole) in order to keep its solar panels locked onto the Sun and the directional antenna pointed at the Earth. Otherwise, both power and communications would have been lost.

A system of gas jets and valves was used periodically to adjust the attitude or position of the spacecraft. Expulsion of nitrogen gas supplied the force for these adjustments during the cruise mode. While the spacecraft was subjected to the heavier disturbances caused by the rocket engine during the midcourse maneuver, the gas jets could not provide enough power to control the attitude of the spacecraft and it was necessary to use deflecting vanes as rudders in the rocket engine exhaust stream for stabilizing purposes.

The attitude control system was activated by CC&S command 60 minutes after launching. It operated first to align the long axis of the spacecraft with the Sun; thus its solar panels would face the Sun. Either the Sun sensors or the three gyroscopes mounted in the pitch (rocking back and forth), yaw (side to side), and roll axes, could activate the gas jet valves during the maneuver, which normally required about 30 minutes to complete.

The spacecraft was allowed a pointing error of 1 degree in order to conserve gas. The system kept the spacecraft swinging through this 1 degree of arc approximately once each 60 minutes. As it neared the limit on either side, the jets fired for approximately $1/50$ of a second to start the swing slowly in the other direction. Thus, Mariner rocked leisurely back and forth throughout its 4-month trip.

Sensitive photomultiplier tubes or electric eyes in the Earth sensor, mounted on the directional antenna, activated the gas jets to roll the spacecraft about the already fixed long axis in order to face the antenna toward the Earth. When the Earth was "acquired," the antenna would then necessarily be oriented in the proper direction. If telemetry revealed that Mariner had accidentally fixed on the Moon, over-ride radio commands from the Earth could restart the orientation sequence.

PROPULSION SYSTEM

The Mariner propulsion system for midcourse trajectory correction employed a rocket engine that weighed 37 pounds with fuel and a nitrogen pressure system, and developed 50 pounds of thrust for a maximum of 57 seconds. The system was suspended within the central portion of the basic hexagonal structure of the spacecraft.

This retro-rocket engine used a type of liquid propellant known as anhydrous hydrazine and it was so delicately controlled that it could burn for as little as $^2/_{10}$ of a second and increase the velocity of the spacecraft from as little as $^7/_{10}$ of a foot per second to as much as 200 feet per second.

The hydrazine fuel was stored in a rubber bladder inside a doorknob-shaped container. At the ignition command, nitrogen gas under 3,000-pound-per-square-inch pressure was forced into the propellant tank through explosively activated valves. The nitrogen then squeezed the rubber bladder, forcing the hydrazine into the combustion chamber.

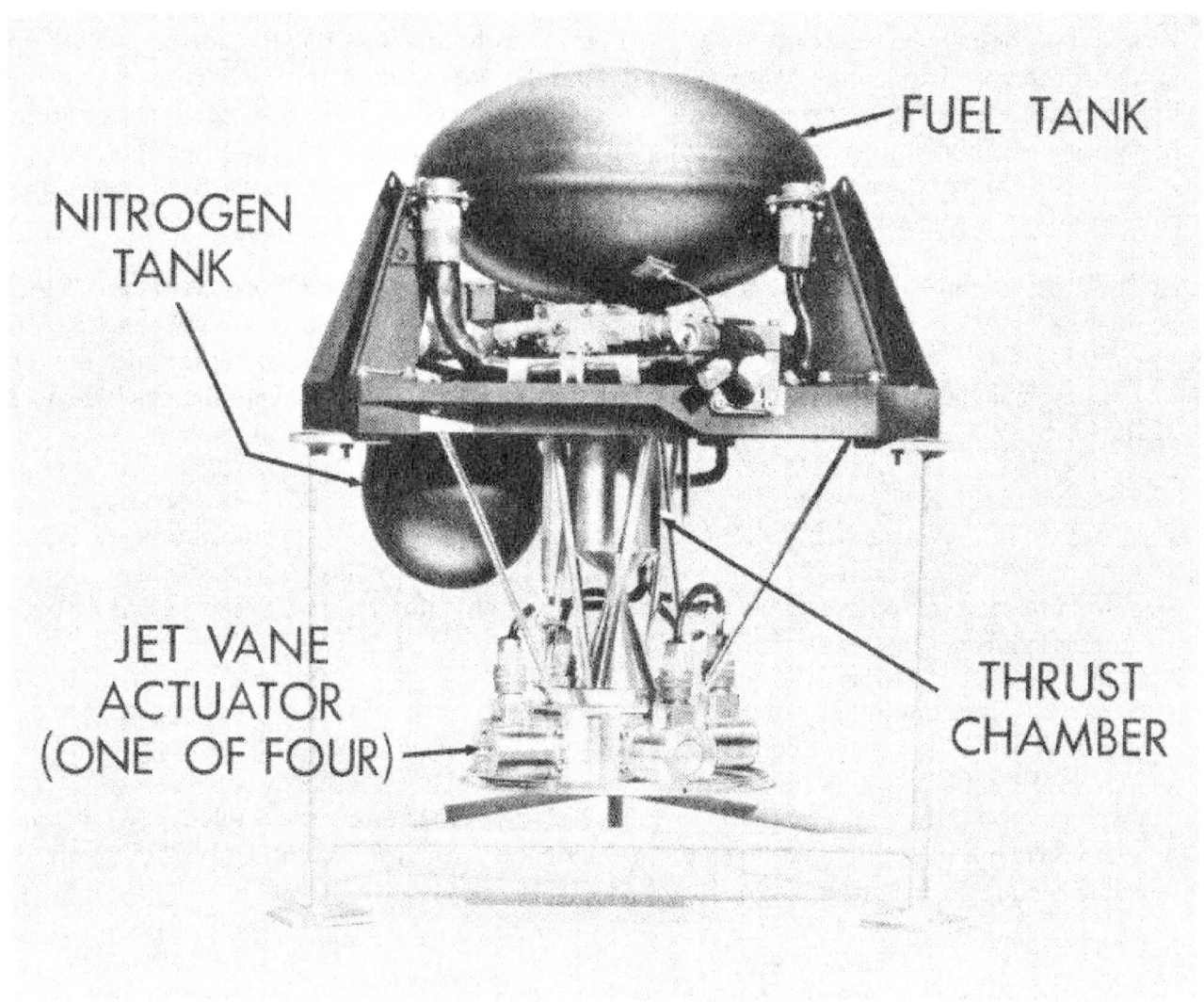

The midcourse propulsion system provides trajectory correction for close approach to Venus.

FUEL TANK

NITROGEN TANK

JET VANE ACTUATOR (ONE OF FOUR)

THRUST CHAMBER

Hydrazine, a monopropellant, requires a starting ignition for proper combustion. In the Mariner system, nitrogen tetroxide starting or "kindling" fluid was injected into the propellant tank by a pressurized cartridge. Aluminum oxide pellets in the tank acted as catalysts to control the speed of combustion of the hydrazine. The burning of the hydrazine was stopped when the flow of nitrogen gas was halted, also by explosively activated valves.

TEMPERATURE CONTROL

Mariner's 129 days in space presented some unique problems in temperature control. Engineers were faced with the necessity of achieving some form of thermal balance so that Mariner would become neither too hot nor too cold in the hostile environment of space.

The spacecraft's temperature control system was made as thermally self-sufficient as possible. Paint patterns, aluminum sheet, thin gold plating, and polished aluminum surfaces reflected and absorbed the proper amount of heat necessary to keep the spacecraft and its subsystems at the proper operating temperatures.

Thermal shields were used to protect the basic hexagon components. The upper shield, constructed of aluminized plastic on a fiberglass panel, protected the top of the basic structure and was designed for maximum immunity to ultraviolet radiation. The lower shield was installed below the hexagon; it was made of aluminum plastic faced with aluminum foil where it was exposed to the blast of the midcourse rocket engine exhaust.

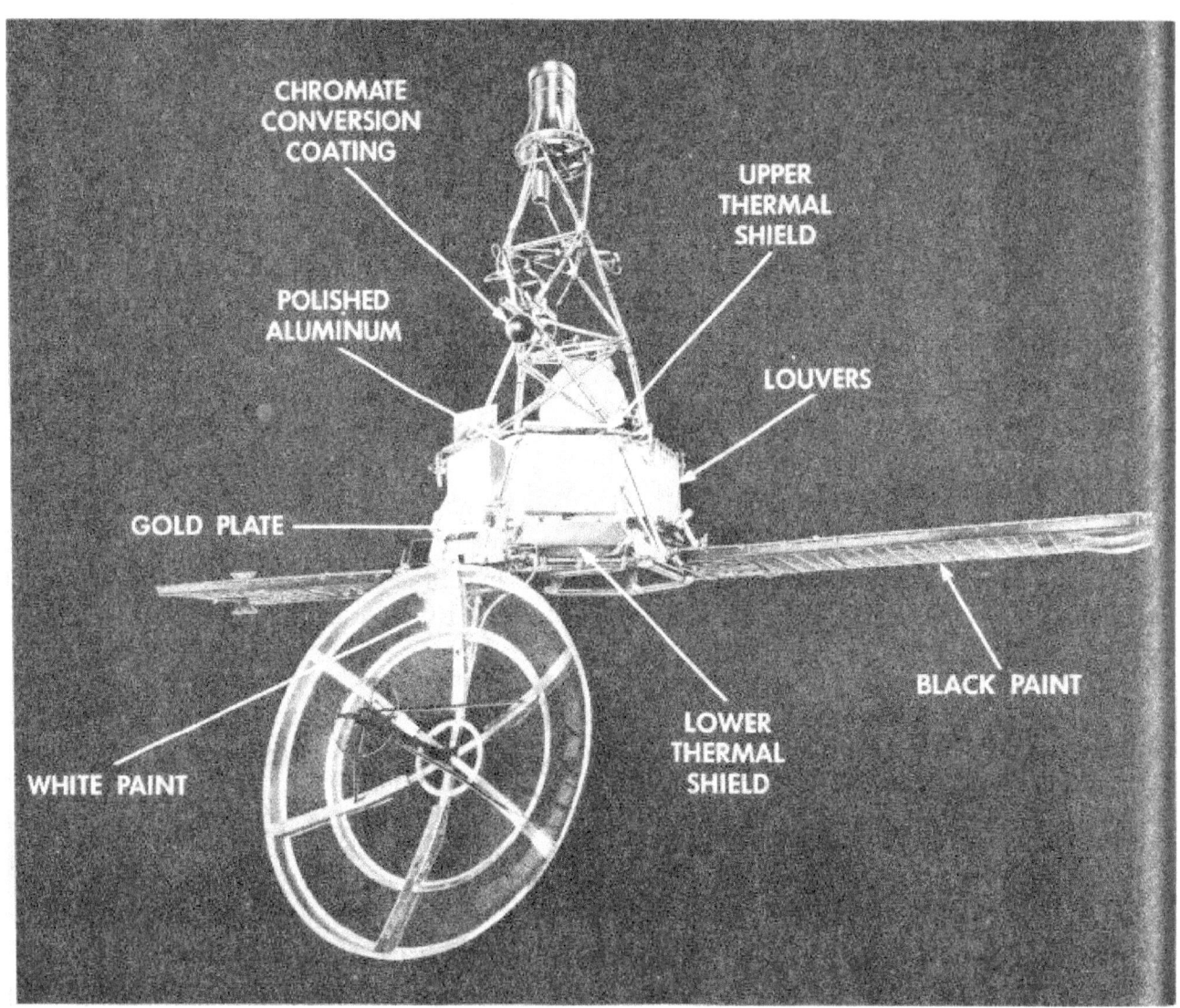

Methods used to control the temperature of the Mariner spacecraft in flight.

CHROMATE CONVERSION COATING

UPPER THERMAL SHIELD

POLISHED ALUMINUM

LOUVERS

GOLD PLATE

BLACK PAINT

LOWER THERMAL SHIELD

The six electronics cases on the hexagon structure were variously treated, depending upon the power of the components contained in each. Those of high power were coated with a good radiating surface of white paint; assemblies of low power were provided with polished aluminum shields to minimize the heat loss.

The case housing the attitude control and CC&S electronics circuits was particularly sensitive because the critical units might fail above 130 degrees F. A special assembly was mounted on the face of this case; it consisted of eight movable, polished aluminum louvers, each actuated by a coiled, temperature-sensitive, bimetallic element. When the temperature rose, the elements acted as springs and opened the louvers. A drop in temperature would close them.

Structures and bracket assemblies external to the basic hexagon were gold plated if made of magnesium, or polished if aluminum. Thus protected, these items became poor thermal radiators as well as poor solar absorbers, making them relatively immune to solar radiation. External cabling was wrapped in aluminized plastic to produce a similar effect.

The solar panels were painted on the shaded side for maximum radiation control properties. Other items were designed so that the internal surfaces were as efficient radiators as possible, thus conserving the spacecraft's heat balance.

THE SCIENTIFIC INSTRUMENTS

Four instruments were operated throughout the cruise and encounter modes of Mariner: a magnetometer, a solar plasma detector, a cosmic dust detector, and a combined charged-particle detector and radiation counter. Two radiometers were used only in the immediate vicinity of Venus.

These instruments are described in detail in Chapter 8.

CHAPTER 4
THE LAUNCH VEHICLE

The motive power of Mariner itself was limited to a trajectory correction rocket engine and an ability, by means of gas jets, to keep its two critical faces pointing at the Sun and the Earth. Therefore, the spacecraft had to be boosted out of the Earth's gravitational field and injected into a flight path accurate enough to allow the trajectory correction system to alter the course to deliver the spacecraft close enough to Venus to be within operating range of the scientific instruments.

The combined Atlas-Agena B booster system which was selected to do the job had a total thrust of about 376,000 pounds. With this power, Atlas-Agena could put 5,000 pounds of payload into a 345-mile orbit, propel 750 pounds on a lunar trajectory, or launch approximately 400 pounds on a planetary mission. This last capability would be taxed to the limit by the 447 pounds of the Mariner spacecraft.

THE ATLAS BOOSTER: POWER OF SIX 707'S

The 360,000 pounds of thrust developed by the Atlas D missile is equivalent to the thrust generated by the engines of six Boeing 707 jet airplanes. All of this awesome power requires a gargantuan amount of fuel: in less than 20 seconds, Atlas consumes more than a propeller-driven, four-engine airplane burns in flying coast-to-coast nonstop.

Photo courtesy of General Dynamics/Astronautics

This military version of the Atlas missile is modified for NASA space flights.

The Atlas missile, as developed by Convair for the Air Force, has a range of 6,300 miles and reaches a top speed of 16,000 miles per hour. The missile has been somewhat modified for use by NASA as a space booster vehicle. Its mission was to lift the second-stage Agena B and the Mariner spacecraft into the proper position and altitude at the right speed so that the Agena could go into Earth orbit, preliminary to the takeoff for interplanetary space.

The Atlas D has two main sections: a body or sustainer section, and a jettisonable aft, or booster engine section. The vehicle measures about 100 feet in length (with military nose cone) and has a diameter of 10 feet at the base. The weight is approximately 275,000 pounds.

No aerodynamic control surfaces such as fins or rudders are used. The Atlas is stabilized and controlled by "gimbaling" or swiveling the engine thrust chambers by means of a hydraulic system. The direction of thrust can thus be altered to control the movements of the missile.

The aft section mounts two 154,500-pound-thrust booster engines and the entire section is jettisoned or separated from the sustainer section after the booster engines burn out. The 60,000-pound-thrust sustainer engine is attached at the center line of the sustainer section. Two 1,000-pound-thrust vernier (fine steering) engines are installed on opposite sides of the tank section in the yaw or side-turn plane.

All three groups of engines operate during the booster phase. Only the sustainer and the vernier engines burn after staging (when the booster engine section is separated from the sustainer section of the missile).

All of the engines use liquid oxygen and a liquid hydrocarbon fuel (RP-1) which is much like kerosene. Dual turbopumps and valves control the flow of these propellants. The booster engine propellants are delivered under pressure to the propellant or combustion chamber, where they are ignited by electroexplosive devices. Each booster thrust chamber can be swiveled a maximum of 5 degrees in pitch (up and down) and yaw (from side to side) about the missile centerline.

The sustainer engine is deflected 3 degrees in pitch and yaw. The outboard vernier engines gimbal to permit pitch and roll movement through 140 degrees of arc, and yaw movement through 20 degrees toward the missile body and 30 degrees outward.

All three groups of engines are started and develop their full rated thrust while the missile is held on the launch pad. After takeoff, the booster engines burn out and are jettisoned. The sustainer engine continues to burn until its thrust is terminated. The swiveled vernier engines provide the final correction in velocity and missile attitude before they are also shut down.

The propellant tank is the basic structure of the forward or sustainer section of the Atlas. It is made of thin stainless steel and is approximately 50 feet long. Internal pressure of helium gas is used to support the tank structure, thus eliminating the need for internal bracing structures, saving considerable weight, and increasing over-all performance of the missile. The helium gas used for this purpose is expanded to the proper pressure by heat from the engines.

Equipment pods on the outside of the sustainer section house the electrical and electronic units and other components of the missile systems.

The Atlas uses a flight programmer, an autopilot, and the gimbaled engine thrust chamber actuators for flight control. The attitude of the vehicle is controlled by the autopilot, which is set for this automatic function before the flight. Guidance commands are furnished by a ground radio guidance system and computer.

The airborne radio inertial guidance system employs two radio beacons which respond to the ground radar. A decoder on board the missile processes the guidance commands.

THE AGENA B: START AND RESTART

Launching Mariner to Venus required a second-stage vehicle capable of driving the spacecraft out of Earth orbit and into a proper flight path to the planet.

Photo courtesy of Lockheed Missiles and Space Company

The Agena B second stage is hoisted to the top of the gantry at AMR.

The Agena B used for this purpose weighs 1,700 pounds, is 60 inches in diameter, and has an over-all length of 25 feet, varying somewhat with the payload. The Agena B fuel tanks are made of 0.080-inch aluminum alloy.

The liquid-burning engine develops more than 16,000 pounds of thrust. The propellants are a form of hydrazine and red fuming nitric acid.

The Agena can be steered to a desired trajectory by swiveling the gimbal-mounted engine on command of the guidance system. The attitude of the vehicle is controlled either by gimbaling the engine or by ejecting gas from pneumatic thrusters.

The Agena has the ability to restart its engine after it has already fired once to reach an Earth orbital speed. This feature makes possible a significant increase in payload and a change of orbital altitude. A velocity meter ends the first and second burns when predetermined velocities have been reached.

After engine cutoff, the major reorientation of the vehicle is achieved through gas jets controlled from an electronic programming device. This system can turn the Agena completely around in orbit, or pitch it down for reentry into the atmosphere. The attitude is controlled by an infrared, heat-sensitive horizon scanner and gyroscopes.

The principal modification to the Agena vehicle for the Mariner II mission was an alteration to the spacecraft-Agena adapter in order to reduce weight.

CHAPTER 5
FLIGHT INTO SPACE

With the Mariner R Project officially activated in the fall of 1961 and the launch vehicles selected, engineers proceeded at full speed to meet the difficult launch schedule.

A preliminary design was adopted in late September, when the scientific experiments to be carried on board were also selected. By October 2, a schedule had been established that would deliver two spacecraft to the assembly building in Pasadena by January 15 and 29, 1962, respectively, with the spares to follow in two weeks.

During the week of November 6, tests were underway to determine problems involved in mating a mock-up of the spacecraft with the Agena shroud and adapter assembly. A thermal control model of the spacecraft had already gone into the small space simulator at JPL for preliminary temperature tests.

MR-1, the first Mariner scheduled for flight, was in assembly immediately after January 8, 1962, and the process was complete by the end of the month, when electrical and magnetic field tests had been started. At the same time, assembly of MR-2 was underway. Work on MR-1 was a week ahead of schedule by the end of the month.

A full-scale temperature control model of the spacecraft went into the large space simulator on February 26. In mid-March, system tests began on both spacecraft and it was decided that the flight hardware would be tested only in the small simulator, with the temperature control model continuing in the large chamber.

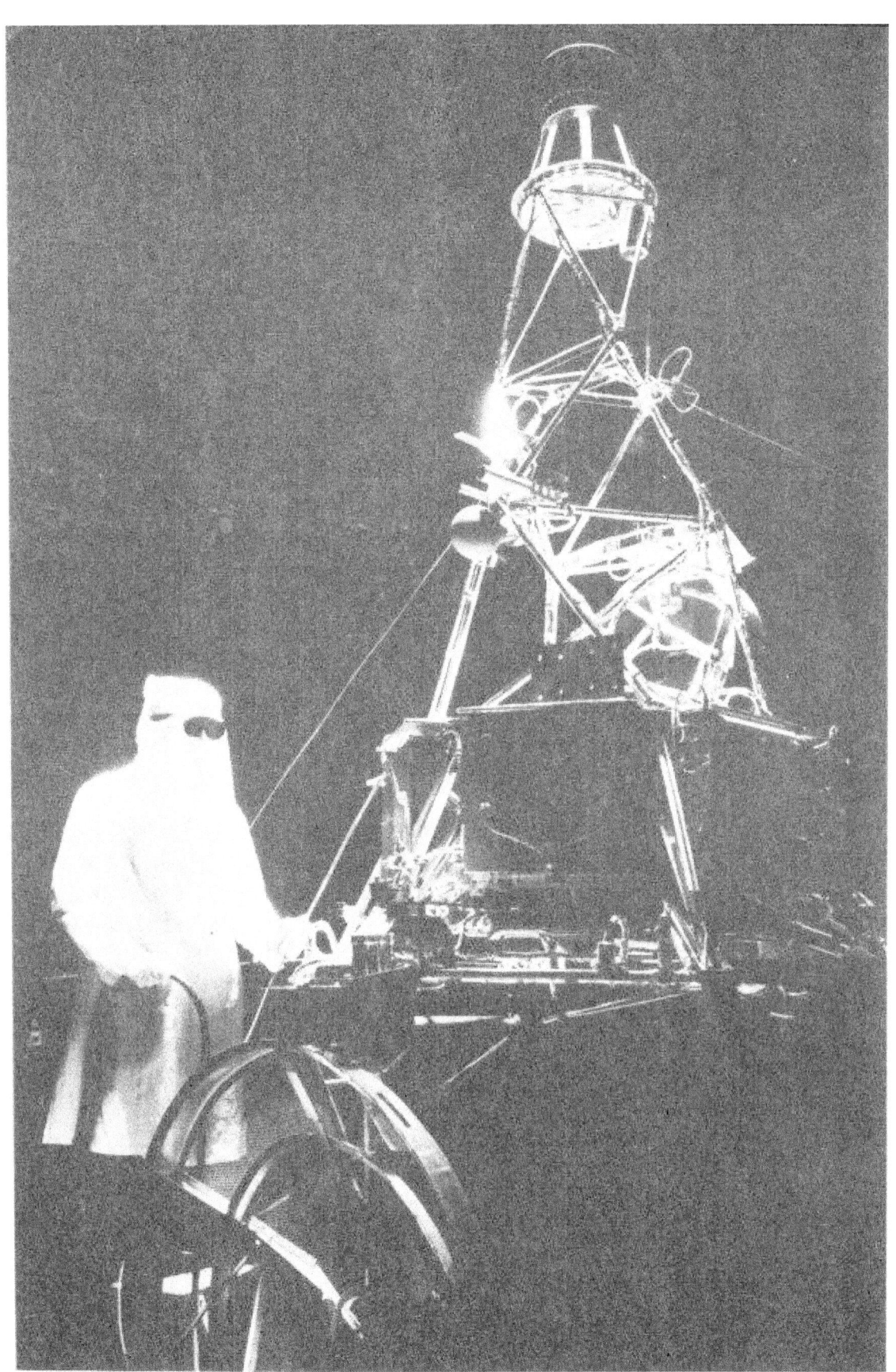

Technician wears hood and protective goggles while working on Mariner spacecraft in Space simulator chamber at Jet Propulsion Laboratory, Pasadena.

On March 26, MR-1 was subjected to full-scale mating tests with the shroud (cover) and the adapter for mounting the spacecraft on top of the Agena. MR-2 was undergoing vibration tests during the week of April 16. By April 30, MR-1 had completed vibration tests and had been mapped for magnetic fields so that, once compensated for, they would not interfere with the magnetometer experiment in space.

A dummy run of MR-1 was conducted on May 7 and the spacecraft, space flight center, and computing equipment were put through a simulated operations test run during the same week.

By May 14, clean-up and final inspection by microscope had begun on MR-1, MR-2, and MR-3 (the latter spacecraft had been assembled from the spares). Soon after, the first two van loads of equipment were shipped to Cape Canaveral. The final system test of MR-1 was completed on May 21 and the test of MR-2 followed during the same week.

During the week of May 28, all three spacecraft and their associated ground support equipment were packed, loaded, and shipped to the Atlantic Missile Range (AMR). At the same time, the Atlas designated to launch MR-1 went aboard a C-133 freight aircraft at San Diego. On the same day, an Air Force order grounded all C-133's for inspection and the plane did not depart until June 9.

By June 11, 1962, the firing dates had been established and both spacecraft were ready for launching. The Atlas booster had already been erected on the launch pad. The dummy run and a joint flight acceptance test were completed on MR-1 during the week of July 2. Final flight preparations and system test of MR-1 and the system test of MR-2 were concluded a week later.

Thus, in 324 days, a new spacecraft project had been activated; the design, assembly, and testing had been completed; and the infinite number of decisions pertaining to launch, AMR Range Operations, deep-space tracking, and data processing activities had been made and implemented.

Venus was approaching the Earth at the end of its 19-month excursion around the Sun. The launch vehicles and Mariners I and II stood ready to go from Canaveral's Launch Complex 12. The events leading to the first close-up look at Venus and intervening space were about to reach their first crisis: a fiery explosion over the Atlantic Ocean.

MARINER I: AN ABORTIVE LAUNCH

After 570 hours of testing, Mariner I was poised on top of the Atlas-Agena launch vehicle during the night of July 20, 1962. The time was right, the Range and the tracking net were standing by, the launch vehicles were ready to cast off the spacecraft for Venus.

Atlas for launching Mariner II arrives at Cape Canaveral in C-133 aircraft.

The countdown was begun at 11:33 p.m., EST, July 20, after several delays because of trouble in the Range Safety Command system. At the time, the launch count stood at T minus 176 minutes—if all went well, 176 minutes until the booster engines were ignited.

Another hold again delayed the count until 12:37 a.m., July 21, when counting was resumed at T minus 165 minutes. The count then proceeded without incident to T minus 79 minutes at 2:20 a.m., when uncertainty over the cause of a blown fuse in the Range Safety circuits caused the operations to be "scrubbed" or cancelled for the night. The next launch attempt was scheduled for July 21-22.

The second launch countdown for Mariner I began shortly before midnight, July 21. Spacecraft power had been turned on at 11:08 p.m., with the launch count at T minus 200 minutes. At T minus 135 minutes, the weather looked good. A 41-minute hold was required at minus 130 minutes (12:17 a.m., July 22) in order to change a noisy component in the ground tracking system.

When counting was resumed at T minus 130 minutes, the clock read 12:48 a.m. A previously scheduled hold was called at T minus 60 minutes, lasting from 1:58 to 2:38 a.m. The good weather still held.

At T minus 80 seconds, power fluctuations in the radio guidance system forced a 34-minute hold. Time was resumed at 4:16 a.m., when the countdown was set back to T minus 5 minutes.

At exactly 4:21.23 a.m., EST, the Atlas thundered to life and lifted off the pad, bearing its Venus-bound load. The boost phase looked good until the Range Safety officer began to notice an unscheduled yaw-left (northeast) maneuver. By 4:25 a.m., it was evident that, if allowed to continue, the vehicle might crash in the North Atlantic shipping lanes or in some inhabited area. Steering commands were being supplied but faulty application of the guidance equations was taking the vehicle far off course.

Finally, at 4:26.16 a.m., after 293 seconds of flight and with just 6 seconds left before separation of the Atlas and Agena—after which the launch vehicle could not be destroyed—a Range Safety officer hit the "destruct" button.

A flash of light illuminated the sky and the choppy Atlantic waters were awash with the glowing death of a space probe. Even as it fluttered down to the sea, however, the radio transponder of the shattered Mariner I continued to transmit for 1 minute and 4 seconds after the destroy command had been sent.

Mariner I did not succumb easily.

MARINER II: A ROLL BEFORE PARKING

Ever since Mariner II had arrived at the Cape on June 4, test teams of all organizations had labored day and night to prepare the spacecraft for launch. The end of their efforts culminated after some 690 hours of test time, both in California and in Florida.

Thirty-five days after Mariner I met its explosive end, the first countdown on Mariner II was underway. At 6:43 p.m., EST, August 25, 1962, time was picked up. The countdown did not proceed far, however. The Atlas crew asked for a hold at T minus 205 minutes (8:39 p.m.) because of stray voltages in the command destruct system caused by a defective Agena battery. After considerable delay, the launch effort was scrubbed at 10:06 p.m.

Two assembly operations and system checkouts are performed separated by a trip to the pad to verify compatibility with the launch vehicle

A complete electronic checkout station in the hangar supports the spacecraft to ensure operability

Mariner takes form as the solar panels are attached and the final hangar checkout operations are performed before the launch.

Wrapped in a dust cover, the spacecraft is transferred from Hangar AE at AMR to the explosive safe area for further tests.

Inside the bunker-like explosive safe area, the powerful midcourse maneuver rocket engine is installed in the center of the spacecraft.

Final assembly and inspection complete, Mariner is "canned" in the nose shroud that will protect it through the Earth's atmosphere and into space.

At the pad, the shrouded spacecraft is lifted past the Atlas ...

... and the Agena.

Twelfth floor: Mariner reaches its mating level.

The spacecraft is eased over to the top of the Agena ...

... and carefully mated to it.

The second launch attempt started at 6:37 p.m., August 26, with the Atlas-Agena B and Mariner II ready on the pad. At 9:52 p.m., T minus 100 minutes, a 40-minute hold was called to replace the Atlas main battery. By 10:37, with 95 minutes to launch, all spacecraft systems were ready to go.

A routine hold at T minus 60 minutes was extended beyond 30 minutes in order to verify the spacecraft battery life expectation. At 11:48 p.m., with the count standing at T minus 55 minutes, the spacecraft, the vehicles, the Range, and the DSIF were all given the green light.

When good launching weather was reported at 12:18 a.m., August 27, just 25 minutes from liftoff, a cautious optimism began to mount in the blockhouse and among the tired crews.

But the tension began to build again. The second prescheduled hold at T minus 5 minutes was extended beyond a half-hour when the radio guidance system had difficulty with ground station power. Counting was "picked up" and the clock continued to move down to 60 seconds before liftoff.

Suddenly, the radio guidance system was in trouble again. Fluctuations showed in its rate beacon signals, and another hold was called. Still another hold for the same reason followed at T minus 50 seconds. This time, at 1:30 a.m., the count was set back to T minus 5 minutes.

One further crisis developed during this hold—only 3 minutes of pre-launch life remained in Atlas' main battery. A quick decision was made to hold the switchover to missile power until T minus 60 seconds to help conserve the life of the battery.

At 1:48 a.m., the count was resumed again at T minus 5 minutes. The long seconds began to drag. Finally, the Convair test director pressed the fire button.

Out on the launch pad, the Atlas engines ignited with a white puff and began to strain against the retaining bolts as 360,000 pounds of thrust began to build up. In a holocaust of noise and flame, the Atlas was released and lifted off the launch pad on a bearing of 106.8 degrees at exactly 1 hour, 53 minutes, 13.927 seconds in the morning of August 27, 1962.

Mariner II was on its way to listen to the music of the spheres.

As the launch vehicle roared up into the night sky, the JPL Launch Checkout Station (DSIF 0) tracked the spacecraft until Mariner disappeared over the horizon. A quick, preliminary evaluation of spacecraft data showed normal readings and Atlas seemed to be flying a true course. The AMR in-flight data transmission and computational operations were being performed as expected. With liftoff out of the way, the launch began to look good.

After the radio signal from the ground guidance system cut off the engines and the booster section was jettisoned, the remaining Atlas forward section, plus the Agena and the spacecraft began to roll. However, it stabilized itself in a normal attitude. Although the Atlas had not gone out of the Range Safety restrictions, it was within just 3 degrees of exceeding the Agena horizon sensor limits, which would have forced another aborted mission.

After the booster separation, the Atlas sustainer and vernier engines continued to burn until they were shut off by radio guidance command. Shortly thereafter, spring-loaded bolts ejected the nose-cone shroud which had protected the spacecraft against frictional heating in the atmosphere. Simultaneously, the gyroscopes in the Agena were started and, at about 1:58 a.m., the Agena and the spacecraft separated from the now-spent Atlas, which was retarded by small retro-rockets and drifted back into the atmosphere, where it was destroyed by friction on reentry.

As the Agena separated from the Atlas booster vehicle, it was programmed to pitch down almost 15 degrees, putting it roughly parallel with the local horizon. Then, following a brief coasting period, the Agena engine ignited at 1:58.53 a.m. and fired until 2:01.12 a.m. Cut-off occurred at a predetermined value of velocity. Both the Agena and the spacecraft had now reached a speed of approximately 18,000 miles per hour and had gone into an Earth orbit at an altitude of 116.19 statute miles.

The second stage and the spacecraft were now in a "parking orbit," which would allow the vehicle to coast out to a point more favorable than Cape Canaveral for blasting off Mariner for Venus.

During the launch, Cape radar had tracked the radar beacon on the Agena, losing it on the horizon at 2:00.53 a.m. Radar stations at Grand Bahama Island, San Salvador, Ascension, the Twin Falls Victory ship, and Pretoria (in South Africa) continued to track down range. Meanwhile, Antigua had "locked on" and tracked the spacecraft's radio transponder and telemetry from 1:58 to 2:08 a.m. when it went over the Antigua horizon.

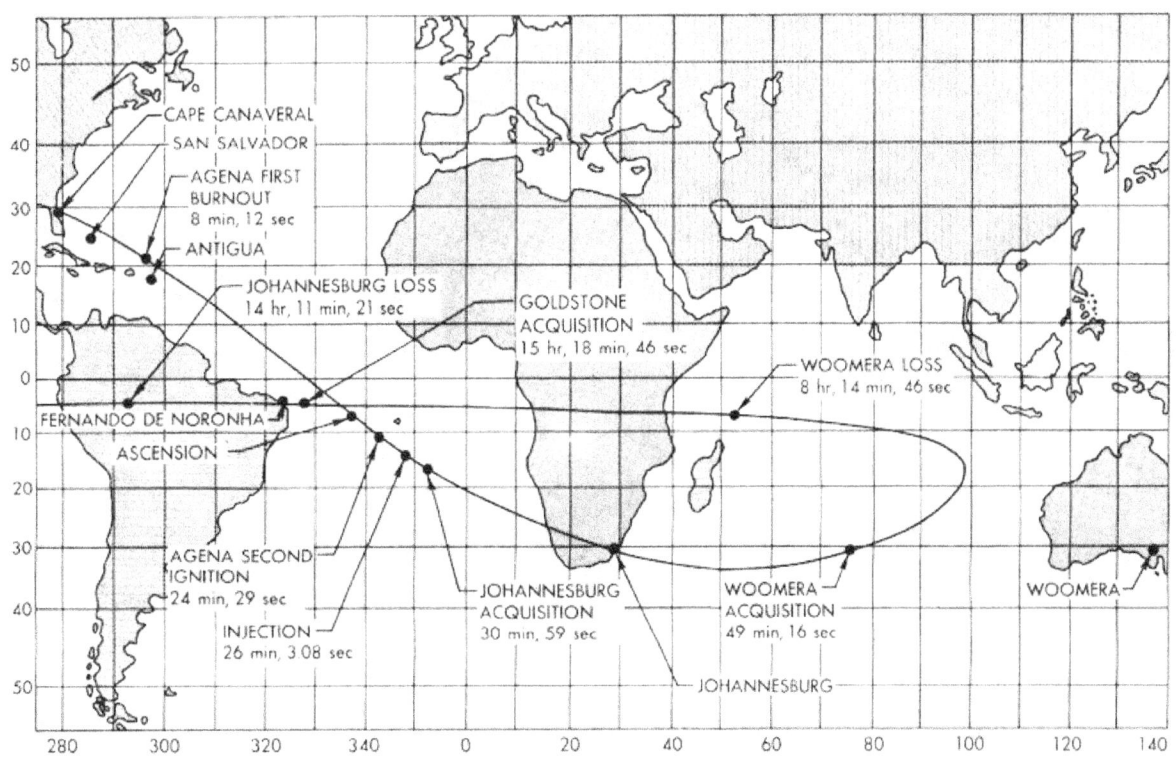

Mariner II is accelerated to Earth-escape velocity and out of orbit near St. Helena. Rotation of earth causes flight path to appear to double back to west over Africa.

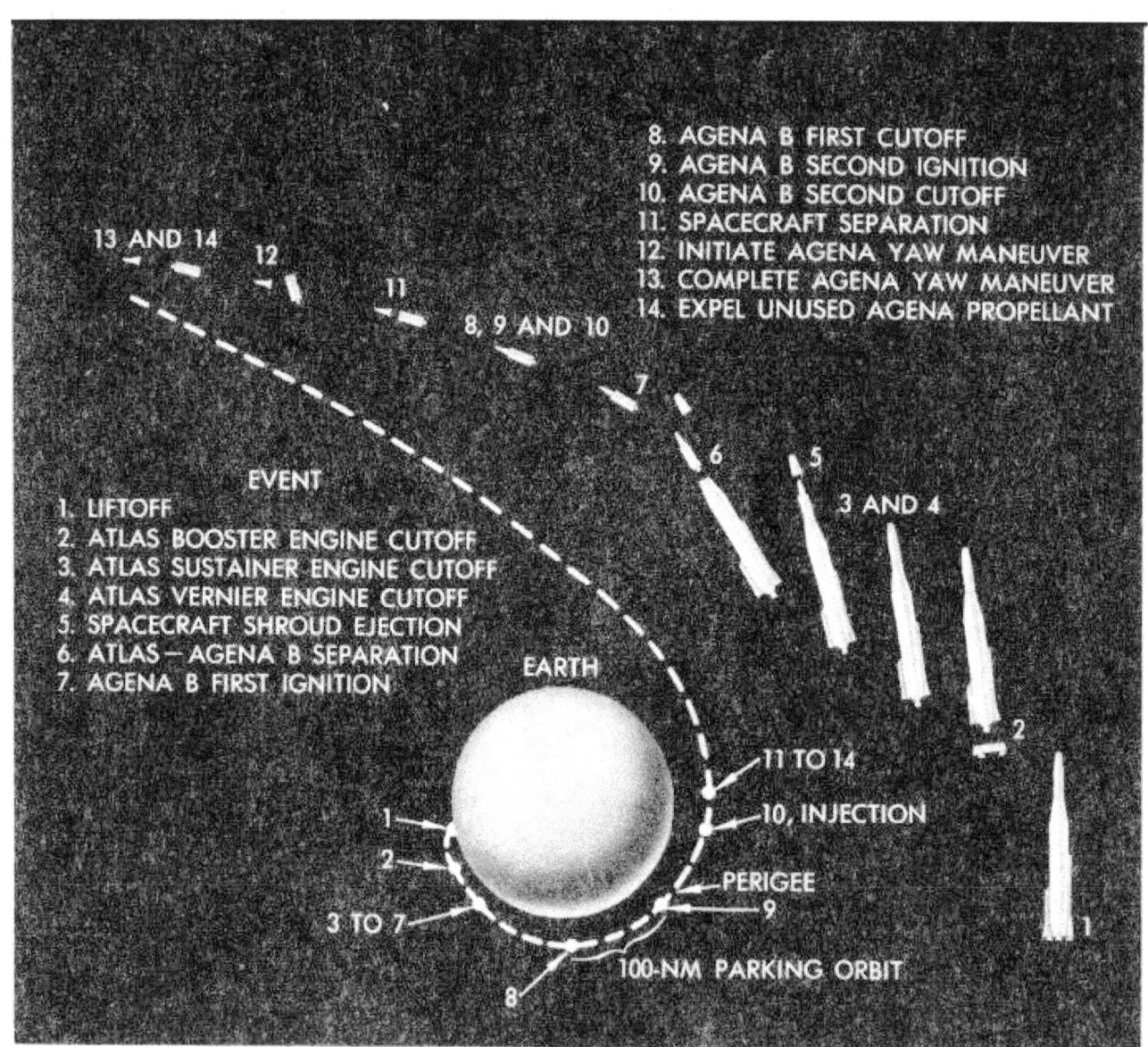

8. AGENA B FIRST CUTOFF
9. AGENA B SECOND IGNITION
10. AGENA B SECOND CUTOFF
11. SPACECRAFT SEPARATION
12. INITIATE AGENA YAW MANEUVER
13. COMPLETE AGENA YAW MANEUVER
14. EXPEL UNUSED AGENA PROPELLANT

13 AND 14   12
11
8, 9 AND 10
7
6   5
3 AND 4

EVENT
1. LIFTOFF
2. ATLAS BOOSTER ENGINE CUTOFF
3. ATLAS SUSTAINER ENGINE CUTOFF
4. ATLAS VERNIER ENGINE CUTOFF
5. SPACECRAFT SHROUD EJECTION
6. ATLAS—AGENA B SEPARATION
7. AGENA B FIRST IGNITION

EARTH

2

1

11 TO 14
10, INJECTION
PERIGEE
9
100-NM PARKING ORBIT

1
2
3 TO 7
8

The sequence of events in the launch phase of the Mariner flight to Venus.

EVENT

1. LIFTOFF

2. ATLAS BOOSTER ENGINE CUTOFF

3. ATLAS SUSTAINER ENGINE CUTOFF

4. ATLAS VERNIER ENGINE CUTOFF

5. SPACECRAFT SHROUD EJECTION

6. ATLAS-AGENA B SEPARATION

7. AGENA B FIRST IGNITION

8. AGENA B FIRST CUTOFF

9. AGENA B SECOND IGNITION

10. AGENA B SECOND CUTOFF

11. SPACECRAFT SEPARATION

12. INITIATE AGENA YAW MANEUVER

13. COMPLETE AGENA YAW MANEUVER

14. EXPEL UNUSED AGENA PROPELLANT

The second coasting period lasted 16.3 minutes, a time determined by the ground guidance computer and transmitted to the Agena during the vernier burning period of Atlas. Then, Agena restarted its engine and fired for a second time. At the end of this firing period, both the Agena and Mariner, still attached, had been injected into a transfer trajectory to Venus at a velocity exceeding that required to escape from the Earth's gravity.

The actual injection into space occurred at 26 minutes 3.08 seconds after liftoff from the Cape (2:19.19 a.m., EST) at a point above 14.873 degrees south latitude and 2.007 degrees west longitude. Thus, Mariner made the break for Venus about 360 miles northeast of St. Helena, 2,500 miles east of the Brazilian coast, and about 900 miles west of Angola on the west African shore.

During injection, the vehicle was being tracked by Ascension, telemetry ship Twin Falls Victory, and Pretoria. Telemetry ship Whiskey secured the spacecraft signal just after injection and tracked until 2:26 a.m. Pretoria began its telemetry track at 2:21 and continued to track for almost two hours, until 4:19 a.m.

Injection velocity was 7.07 miles per second or 25,420 miles per hour, just beyond Earth-escape speed. The distance at the time of injection from Canaveral's Launch Complex 12 was 4,081.3 miles.

The Agena and Mariner flew the escape path together for another two minutes after injection before they were separated at 2:21 a.m. Agena then performed a 140-degree yaw or retro-turn maneuver by expelling unused propellants. The purpose was to prevent the unsterilized Agena from possibly hitting the planet, and from following Mariner too closely and perhaps disturbing its instruments.

Now, Mariner II was flying alone and clear. Ahead lay a journey of 109 days and more than 180 million miles.

ORIENTATION AND MIDCOURSE MANEUVER

As Mariner II headed into space, the Deep Space Instrumentation Facility (DSIF) network began to track the spacecraft. At 2:23.59 a.m., DSIF 5 at Johannesburg, aided by the Mobile Tracking Station, installed in vans in the vicinity, was "looking" at the spacecraft, just four minutes after injection.

Johannesburg was able to track Mariner until 4:04 p.m. because, as the trajectory took Mariner almost radially away from the Earth, our planet began in effect to turn away from under the spacecraft. On an Earth map, because of its course and the rotation of the Earth, Mariner II appeared to describe a great arc over the Indian Ocean far to the west of Australia, then to turn north and west and to proceed straight west over south-central Africa, across the Atlantic, and over the Amazon Basin of northern South America. Johannesburg finally lost track at a point over the middle of South America.

While swinging over the Indian Ocean on its first pass, the spacecraft was acquired by Woomera's DSIF 4 at 2:42.30 a.m., and tracked until 8:08 a.m., when Mariner was passing just to the north of Madagascar on a westerly course. Goldstone did not acquire the spacecraft until it was approaching the east coast of South America at 3:12 p.m., August 27.

With Mariner slowly tumbling in free space, it was now necessary to initiate a series of events to place the spacecraft in the proper flight position. At 2:27 a.m., 44 minutes after launch, the Mariner Central Computer and Sequencer (CC&S) on board the spacecraft issued a command for explosively activated pin pullers to release the solar panels and the radiometer dish from their launch-secured positions. At 2:53, 60 minutes after liftoff, the attitude control system was turned on and the Sun orientation sequence began with the extension of the directional antenna to a preset angle of 72 degrees.

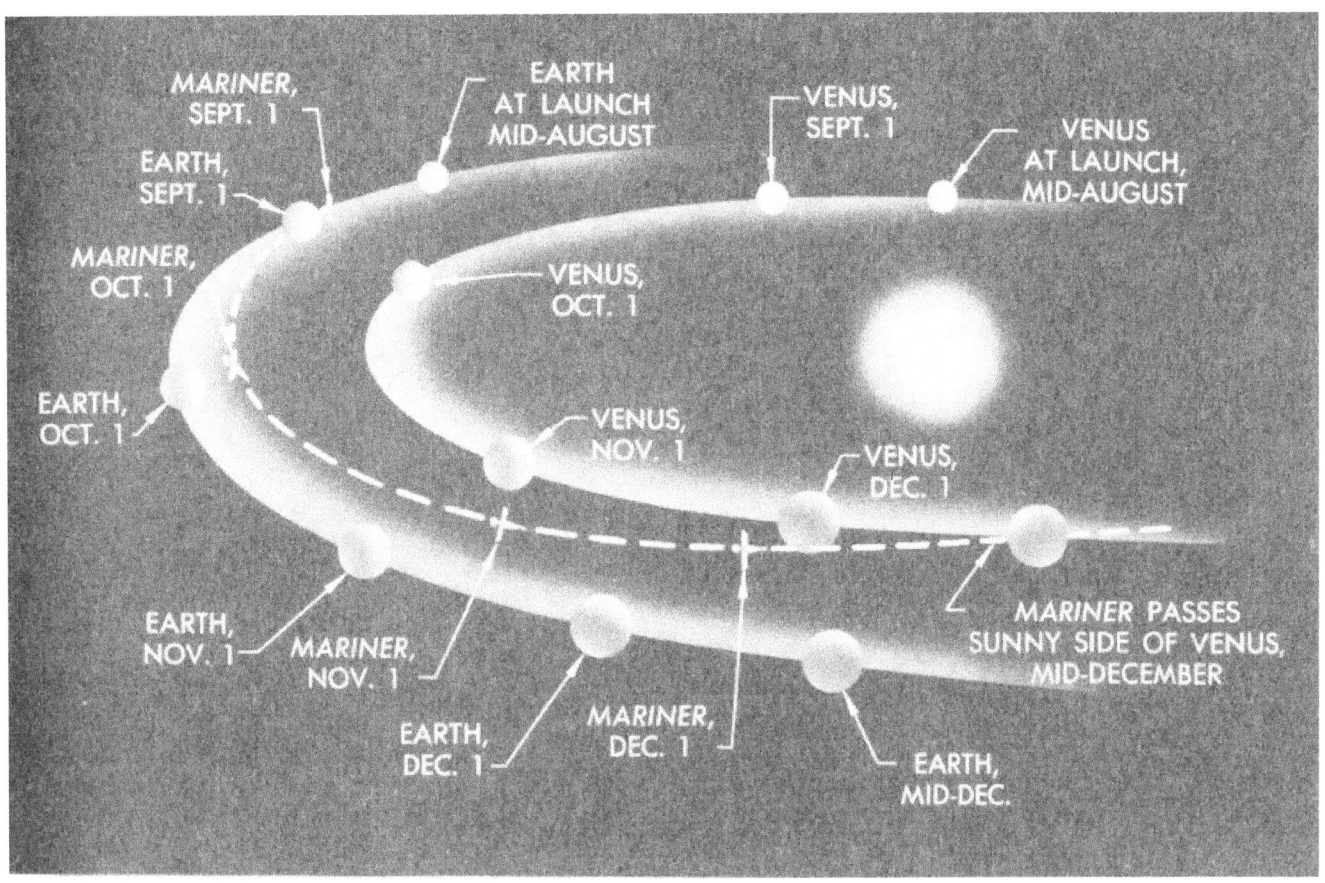

Mariner II was launched while Venus was far behind the Earth. During the 109-day flight, Venus overtook and passed the Earth. It rendezvoused with the spacecraft at a point about 36,000,000 miles from the Earth.

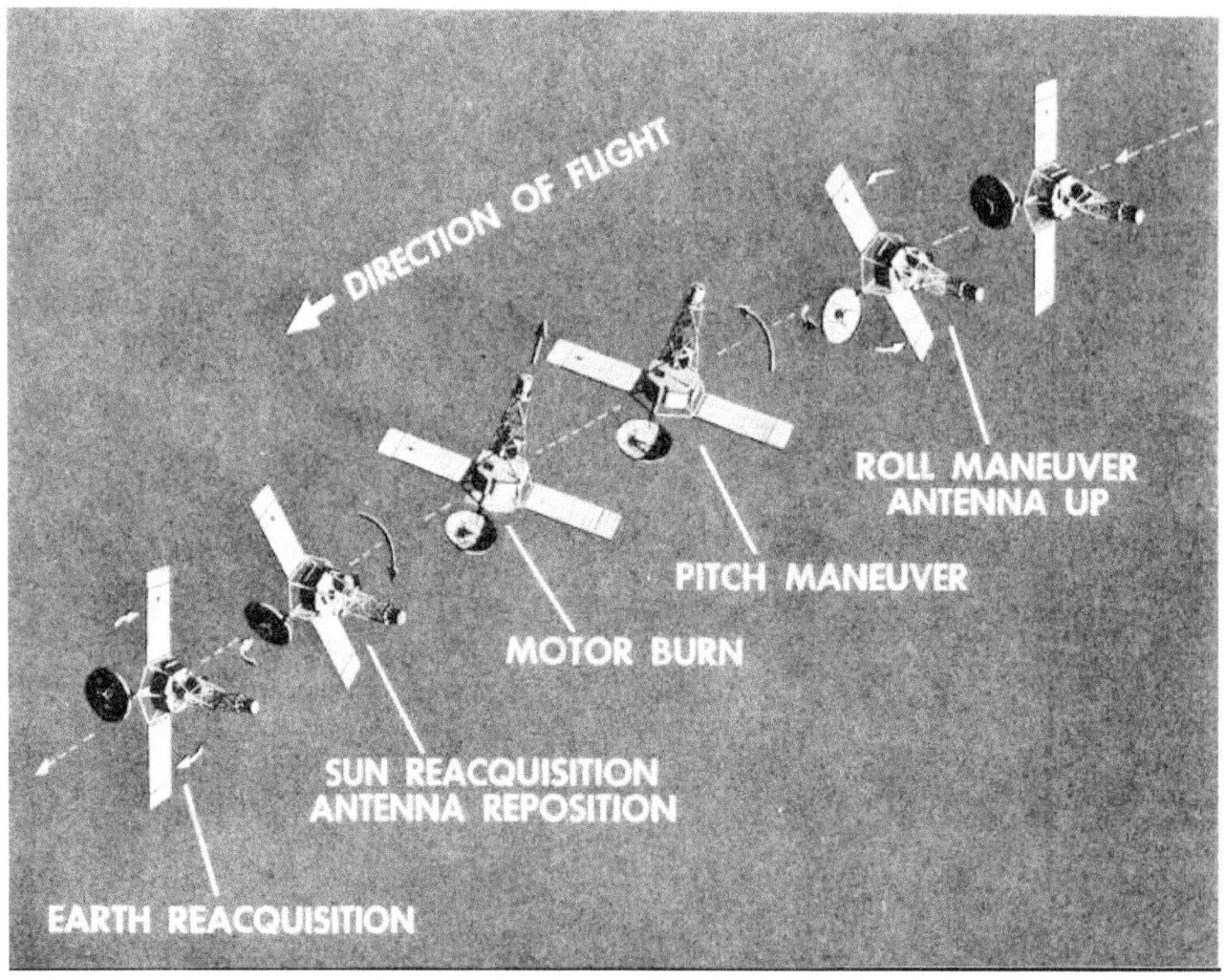

During the midcourse maneuver, the trajectory of Mariner II was corrected so that the spacecraft would approach within 21,598 miles of Venus.

ROLL MANEUVER ANTENNA UP

PITCH MANEUVER

MOTOR BURN

SUN REAQUISITION ANTENNA REPOSITION

EARTH REAQUISITION

The Sun sensors then activated the gas jets and moved the spacecraft about until the roll or long axis was pointed at the Sun. This maneuver required only 2½ minutes after the CC&S

issued the command. The solar panel power output of 195 watts was somewhat higher than anticipated, as were the spacecraft temperatures, which decreased and stabilized six hours after the spacecraft oriented itself on the Sun.

On August 29, a command from Johannesburg turned on the cruise scientific experiments, including all the instruments except the two radiometers. The rate of data transmission was then observed to decrease as planned and the data conditioning system was functioning normally.

For seven days, no attempt was made to orient the spacecraft with respect to the Earth because the Earth sensors were too sensitive to operate properly at such a close range. On September 3, the CC&S initiated the Earth acquisition sequence. The gyroscopes were turned on, the cruise scientific instruments were temporarily switched off, and a search for the Earth began about the roll axis of the spacecraft.

During this maneuver, the long axis of the spacecraft was held steady in a position pointing at the Sun and the gas jets rolled the spacecraft around this axis until the sensors, mounted in the directional antenna, could "see" the Earth. Apparently, the Earth sensor was already viewing the Earth because the transmitter output immediately switched from the omni- to the directional antenna, indicating that no search was necessary.

However, the initial brightness reading from the Earth sensor was 38, an intensity that might be expected if the spacecraft were locked onto the Moon instead of the Earth. As a result, the midcourse maneuver was delayed until verification of Earth lock was obtained.

Mariner's injection into the Venus trajectory yielded a predicted miss of 233,000 miles in front of the planet, well within the normal miss pattern expected as a result of the launch. Because the spacecraft was designed to cross the orbit of Venus behind the planet and pass between it and the Sun, it was necessary to correct the trajectory to an approximate 8,000- to 40,000-mile "fly-by" so the scientific instruments could operate within their design ranges.

After comparison of the actual flight path with that required for a proper near-miss, the necessary roll, pitch, and motor-burn commands were generated by the JPL computers. When, on September 4, it had been established that the spacecraft was indeed oriented on the Earth and not the Moon, a set of three commands was transmitted to the spacecraft from Goldstone, to be stored in the electronic "memory unit" until the start command was sent.

At 1:30 p.m., PST, the first commands were transmitted: a 9.33-degree roll turn, a 139.83-degree pitch turn, and a motor-burn command to produce a 69.5-mile-per-hour velocity change.

At 2:39 p.m., a fourth command was sent to switch from the directional antenna to the omni-antenna. Finally, a command went out instructing the spacecraft to proceed with the now "memorized" maneuver program.

Mariner then turned off the Earth and Sun sensors, moved the directional antenna out of the path of the rocket exhaust stream, and executed a 9.33-degree roll turn in 51 seconds.

Next, the pitch turn was completed in 13¼ minutes, turning the spacecraft almost completely around so the motor nozzle would point in the correct direction when fired.

The spacecraft was stabilized and the roll and pitch turns controlled by gyroscopes, which signalled the attitude control system the rate of correction for comparison with the already computed values.

With the solar panels no longer directly oriented on the Sun, the battery began to share the power demand and finally carried the entire load until the spacecraft had again been oriented on the Sun.

At the proper time, the motor—controlled by the CC&S—ignited and burned for 27.8 seconds, while the spacecraft's acceleration was compared with the predetermined values by the accelerometer. During this period, when the gas jets could not operate properly, the spacecraft was stabilized by movable vanes or rudders in the exhaust of the midcourse motor.

The velocity added by the midcourse motor resulted in a decrease of the relative speed of the spacecraft with respect to the Earth by 59 miles per hour (from 6,748 to 6,689 miles per hour), while the speed relative to the Sun increased by 45 miles per hour (from 60,117 to 60,162 miles per hour).

This apparently paradoxical condition occurred because, in order to intercept Venus, Mariner had been launched in a direction opposite to the Earth's course around the Sun. The midcourse maneuver turned the spacecraft around and slowed its travel away from the Earth while allowing it to increase its speed around the Sun in the direction of the Earth's orbit. Gradually, then, the spacecraft would begin to fall in toward the Sun while moving in the same direction as the Earth, catching and passing the Earth on the 65th day and intersecting Venus' orbit on the 109th day.

At the time of the midcourse maneuver, the spacecraft was travelling slightly inside the Earth's orbit by 70,000 miles, and was behind the Earth by 1,492,500 miles.

THE LONG CRUISE

After its completion of the midcourse maneuver, Mariner reoriented itself on the Sun in 7 minutes and on the Earth in about 30 minutes. During the midcourse maneuver, the omnidirectional antenna was used; now, with the maneuver completed, the directional antenna was switched back in for the duration of the mission.

Ever since the spacecraft had left the parking orbit near the Earth and been injected into the Venus trajectory, the Space Flight Operations Center back in Pasadena had been the nerve center of the mission. Telemetered data had been coming in from the DSIF stations on a 24-hour schedule. During the cruise phase, from September 5 to December 7, a total of 16 orbit computations were made to perfect the planet encounter prediction. On December 7, the first noticeable Venus-caused effects on Mariner's trajectory were observed, causing a definite deviation of the spacecraft's flight path.

On September 8, at 12:50 p.m., EST, the spacecraft lost its attitude control, which caused the power serving the scientific instruments to switch off and the gyroscopes to switch on automatically for approximately three minutes, after which normal operation was resumed. The cause was not apparent but the chances of a strike by some small space object seemed good.

As a result of this event, a significant difference in the apparent brightness reading of the Earth sensor was noted. This sensor had been causing concern for some time because its readings had decreased to almost zero. Further decrease, if actually caused by the instrument and not by the telemetry sensing elements, could result in loss of Earth lock and the failure of radio contact.

After the incident of September 8, the Earth sensor brightness reading increased from 6 to 63, a normal indication for that day. Thereafter, this measurement decreased in an expected manner as the spacecraft increased its distance from the Earth.

Mariner II was now embarked on the long cruise. On September 12, the distance from the Earth was 2,678,960 miles and the spacecraft speed relative to the Earth was 6,497 miles per hour. Mariner was accelerating its speed as the Sun's gravity began to exert a stronger pull than the Earth's. On October 3, Mariner was nearly 6 million miles out and moving at 6,823 miles per hour relative to the Earth. A total of 55,600,000 miles had been covered to that point.

Considerable anxiety had developed at JPL when Mariner's Earth sensor reading had fallen off so markedly. This situation was relieved by the unexplained return to normal on September 8, although the day-to-day change in the brightness number was watched closely. The apparent ability of the spacecraft to recover its former performance after the loss of attitude control on September 8 and again on September 29 was an encouraging sign.

Another disturbing event occurred on October 31, when the output from one solar panel deteriorated abruptly. The entire power load was thrown on the other panel, which was then dangerously near its maximum rated output. To alleviate this situation, the cruise scientific instruments were turned off. A week later, the malfunctioning panel returned to normal operation and the science instruments were again turned on. Although the trouble had cleared temporarily, it developed again on November 15 and never again corrected itself. The diagnosis was a partial short circuit between one string of solar cells and the panel frame, but by now the spacecraft was close enough to the Sun so that one panel supplied enough power.

By October 24, the spacecraft was 10,030,000 miles from the Earth and was moving at 10,547 miles per hour relative to the Earth. The distance from Venus was now 21,266,000 miles.

October 30 was the 65th day of the mission and at 5 a.m., PST, Mariner overtook and passed the Earth at a distance of 11,500,000 miles. Since the spacecraft's direction of travel had, in effect, been reversed by the midcourse maneuver, it had been gaining on the Earth in the direction of its orbit, although constantly falling away from the Earth in the direction of the Sun.

The point of equal distance between the Earth and Venus was passed on November 6, when Mariner was 13,900,000 miles from both planets and travelling at 13,843 miles per hour relative to the Earth. As November wore on, hope for a successful mission began to mount. Using tracking data rather than assumptions of standard midcourse performance, the Venus miss distance had now been revised to about 21,000 miles and encounter was predicted for December 14. But the DSIF tracking crews, the space flight and computer operators, and the management staff could not yet relax. The elation following the successful trajectory correction maneuver on September 4 had given way alternately to discouragement and guarded optimism.

Four telemetry measurements were lost on December 9 and never returned to normal. They measured the angle of the antenna hinge, the fuel tank pressure, and the nitrogen pressure in the midcourse and attitude control systems. A blown fuse, designed to protect the data encoder from short circuits in the sensors, was suspected. However, these channels could not affect spacecraft operation and Mariner continued to perform normally.

The rising temperatures recorded on the spacecraft were more serious. Only the solar panels were displaying expected temperature readings; some of the others were as much as 75 degrees above the values predicted for Venus encounter. The heat increase became more rapid after November 20. By December 12, six of the temperature sensors had reached their upper limits. It was feared that the failure point of the equipment might be exceeded.

The CC&S performed without incident until just before encounter, when, for the first time, it failed to yield certain pulses. JPL engineers were worried about the starting of the encounter sequence, due the next day, although they knew that Earth-based radio could send these commands, if necessary.

On December 12, with the climax of the mission near, the spacecraft was 34,218,000 miles from the Earth, with a speed away from the Earth of 35,790 miles per hour, a Sun-relative speed of 83,900 miles per hour.

Only 635,525 miles from Venus at this point, Mariner II was closing fast on the cloud-shrouded planet. But it was a hot spacecraft that was carrying its load of inquisitive instruments to the historic encounter.

On its 109th day of travel, Mariner had approached Venus in a precarious condition. Seven of the over-heated temperature sensors had reached their upper telemetry limits. The Earth-sensor brightness reading stood at 3 (0 was the nominal threshold) and was dropping. Some 149 watts of power were being consumed out of the 165 watts still available from the crippled solar panels.

At JPL's Space Flight Operations Center, there was reason to believe that the ailing CC&S might not command the spacecraft into its encounter sequence at the proper time. Twelve hours before encounter, these fears were verified.

Quickly, the emergency Earth-originated command was prepared for transmission. At 5:35 a.m., PST, a radio signal went out from Goldstone's Echo Station. Thirty-six million miles away, Mariner II responded to the tiny pulse of energy from the Earth and began its encounter sequence.

After Mariner had "acknowledged" receipt of the command from the Earth, the spacecraft switched into the encounter sequence as engineering data were turned off and the radiometers began their scanning motion, taking up-and-down readings across the face of the planet. As throughout the long cruise, the four experiments monitoring the magnetic fields, cosmic dust, charged particles, and solar plasma experiments continued to operate.

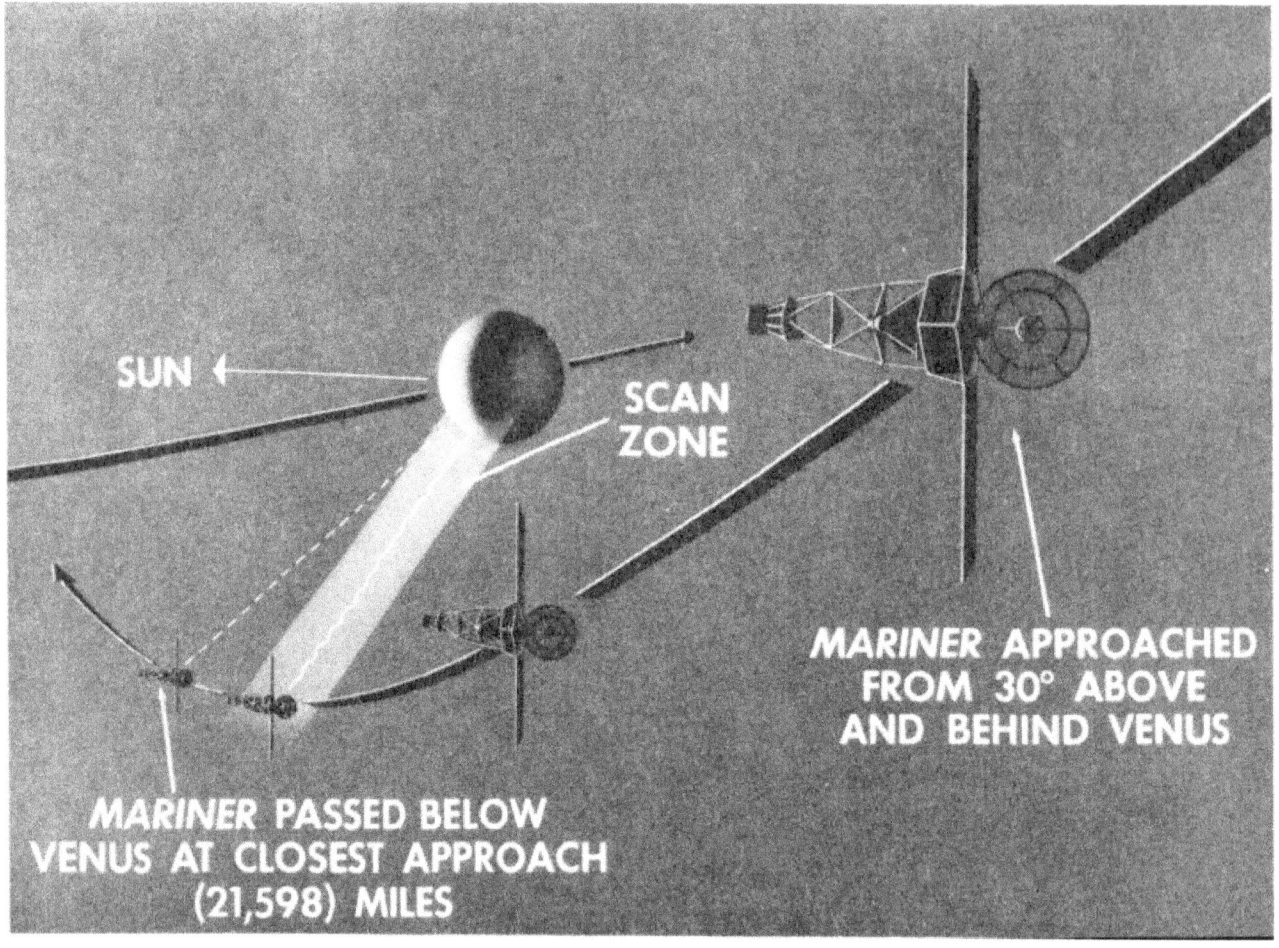

Mariner II approached Venus from the dark side, crossed between the planet and the Sun while making three radiometer scans of the disk.

As Mariner approached Venus on its night side, it was travelling about 88,400 miles per hour with respect to the Sun. At the point of closest approach, at 11:59.28 a.m., PST, the distance from the planet was 21,598 miles.

During encounter with Venus, three scans were made: one on the dark side, one across the terminator dividing dark and sunlit sides, and one on the sunlit side. Although the scan went slightly beyond the edge of the planet, the operation proceeded smoothly and good data were received on the Earth.

With encounter completed, the cruise condition was reestablished by radio command from the Earth and the spacecraft returned to transmitting engineering data, together with the continuing readings of the four cruise scientific experiments.

After approaching closer to a planet and making more meaningful scientific measurements than any man-made space probe, Mariner II continued on into an orbit around the Sun.

December 27, 13 days after Venus encounter, marked the perihelion, or point of Mariner's closest approach to the Sun: 65,505,935 miles. The Sun-related speed was 89,442 miles per hour. As Mariner began to pull away from the Sun in the following months, its Sun-referenced speed would decrease.

Data were still being received during these final days and the Earth and Sun lock were still being maintained. Although the antenna hinge angle was no longer being automatically readjusted by the spacecraft, commands were sent from the Earth in an attempt to keep the antenna pointed at the Earth, even if the Earth sensor were no longer operating properly.

At 2 a.m., EST, January 3, 1963, 20 days after passing Venus, Mariner finished transmitting 30 minutes of telemetry data to Johannesburg and the station shut down its operation. When Woomera's DSIF 4 later made a normal search for the spacecraft signal, it could not be found. Goldstone also searched in vain for the spacecraft transmissions, but apparently Mariner's voice had at last died, although the spacecraft would go into an eternal orbit around the Sun.

It was estimated that Mariner's aphelion (farthest point out) in its orbit around the Sun would occur on June 18, 1963, at a distance of 113,813,087 miles. Maximum distance from the Earth would be 98,063,599 miles on March 30, 1963; closest approach to the Earth: 25,765,717 miles on September 27, 1963.

THE RECORD OF MARINER

The performance record of Mariner II exceeded that of any spacecraft previously launched from Earth:

- It performed the first and most distant trajectory-correcting maneuver in deep space, firing a rocket motor at the greatest distance from the Earth: 1,492,000 miles (September 4, 1962).

- The spacecraft transmitted continuously for four months, sending back to the Earth some 90 million bits of information while using only 3 watts of transmitted power.

- Useful telemetry measurements were made at another record distance from the Earth: 53.9 million miles (January 3, 1963).

- Mariner II was the first spacecraft to operate in the immediate vicinity of another planet and return useful scientific information to the Earth: approximately 21,598 miles from Venus (December 14, 1962).

- Measurements were made closest to the Sun: 65.3 million miles away (December 27, 1962).

- Mariner's communication system operated for the longest continuous period in interplanetary space: 129 days (August 27, 1962, to January 3, 1963).

- Mariner achieved the longest continuous operation of a spacecraft attitude-stabilization system in space, and at a greater distance from the Earth than any previous spacecraft: 129 days (August 27, 1962, to January 3, 1963), at 53.9 million miles from the Earth.

CHAPTER 6
THE TRACKING NETWORK

Thirty-six million miles separated the Earth from Venus at encounter. Communicating with Mariner II and tracking it out to this distance, and beyond, represented a tremendous extension of man's ability to probe interplanetary space.

The problem involved:

1. The establishment of the spacecraft's velocity and position relative to the Earth, Venus, and the Sun with high precision.

2. The transmission of commands to activate spacecraft maneuvers.

3. The reception of readable spacecraft engineering and scientific data from the far-ranging Mariner.

The tracking network had to contend with many radio noise sources: the noise from the solar system and from extragalactic origins; noise originating from the Earth and its atmosphere; and the inherent interference originating in the receiving equipment. These problems were solved by using advanced high-gain antennas and ultra-stable, extremely sensitive receiving equipment.

DEEP SPACE INSTRUMENTATION FACILITY

The National Aeronautics and Space Administration has constructed a network of deep-space tracking stations for lunar and planetary exploration missions. In order to provide continuous, 24-hour coverage, three stations were built, approximately 120 degrees of longitude apart, around the world: at Goldstone in the California desert, near Johannesburg in South Africa, and at Woomera in the south-central Australian desert.

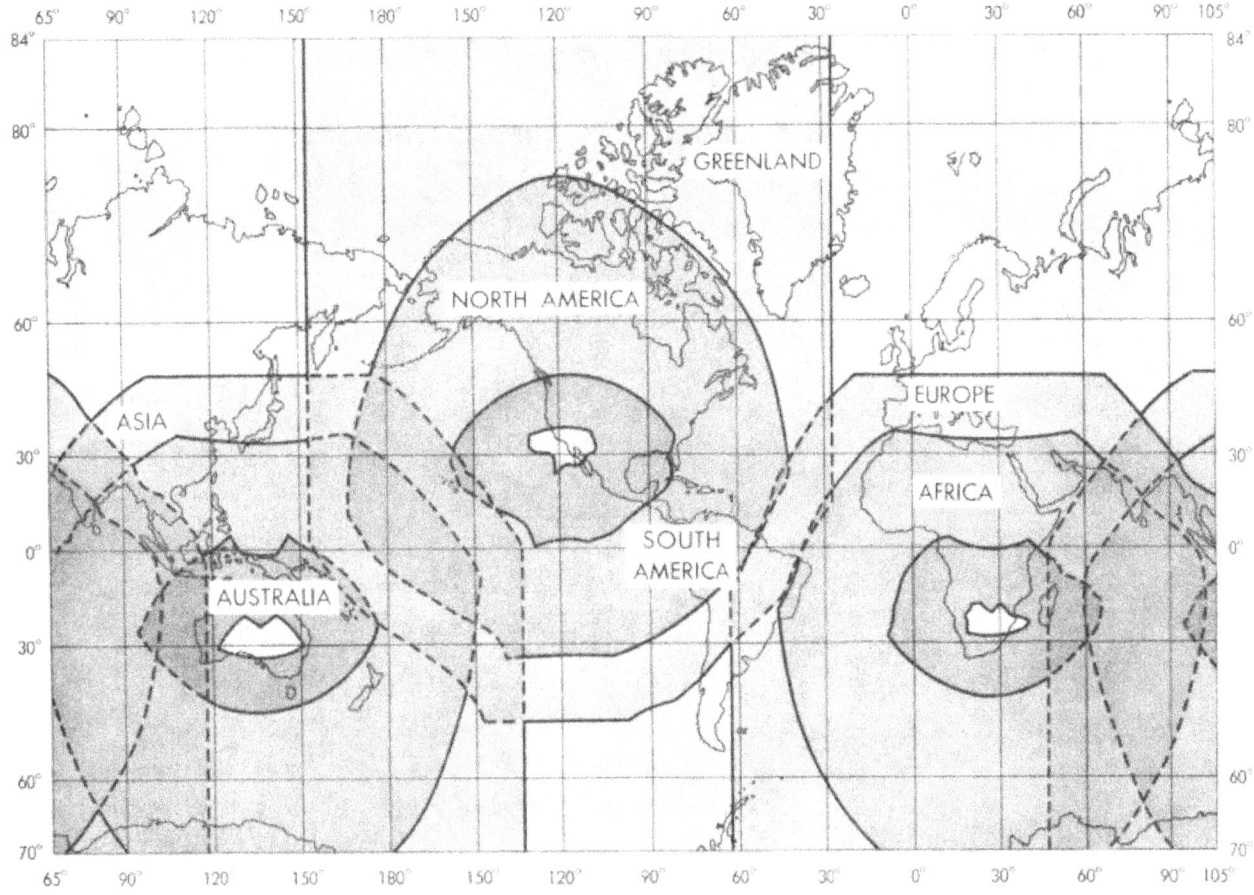

The three tracking stations of the Deep Space Instrumentation Facility are located around the world so as to provide continuous flight coverage.

These stations are the basic elements of the Deep Space Instrumentation Facility (DSIF). In addition, a mobile tracking station installed in vans is used near the point of injection of a spacecraft into an Earth-escape trajectory to assist the permanent stations in finding the spacecraft and to acquire tracking data. The control point for the DSIF net is located at JPL in Pasadena, California (see Table 1).

The Jet Propulsion Laboratory has the responsibility for the technical direction of the entire DSIF net and operates the Goldstone facilities with assistance from the Bendix Corporation as a subcontractor. The overseas stations are staffed and operated by agencies of the Republic of South Africa and the Commonwealth of Australia.

The DSIF net tracks the position and velocity of U.S. deep-space probes, issues commands to direct the spacecraft in flight, receives engineering and scientific data from the probes, and automatically relays the data to JPL in Pasadena, where it is processed by computers and interpreted. (In the tracking operation, a signal is transmitted to the spacecraft, where it is received and processed in a transponder, which then sends the signal back to the

Earth. The change in frequency, known as the doppler effect, involved in this operation enables engineers to determine the velocity at which the spacecraft is moving.)

Table 1. Deep Space Instrumentation Facility Stations

Station	Location	Equipment	Functions
DSIF 1 (Mobile Tracking Station)	Near point of injection of spacecraft into Earth-escape trajectory	10-ft antenna 25-w, 890-mc transmitter	Fast tracking for acquisition of spacecraft
Goldstone:	California		
Pioneer Site (DSIF 2)		85-ft polar-mount antenna; Cassegrain feed; maser and parametric amplifier	Reception of telemetry Tracking spacecraft
Echo Site (DSIF 3)		85-ft polar-mount antenna; parametric amplifier 10-kw, 890-mc transmitter	Transmission of commands Tracking spacecraft Stand-by reception
Venus Site		85-ft radar-type antenna	Advanced radar astronomy Communications research
DSIF 4	Woomera, Australia	85-ft polar-mount antenna; parametric amplifier	Reception of telemetry Tracking spacecraft
DSIF 5	Johannesburg, South Africa	85-ft polar-mount antenna; parametric amplifier 10-kw, 890-mc transmitter	Reception of telemetry Tracking spacecraft Transmission of commands

The stations are equipped with receiving and tracking instruments so sensitive that engineers estimate that they can detect radio-frequency energy equivalent to that radiated by a 1-watt light bulb at a distance of approximately 75 to 80 million miles. Such energy received at the antenna would measure about 0.00000000000000000002 watt (2×10^{-20}).

The amount of power received at the antenna during Mariner's encounter with Venus has been calculated at about 0.000000000000000001 of a watt (1×10^{-18}). If a 100 percent efficient storage battery were charged with this amount of energy for some 30 billion years, the battery would then have stored enough energy to light an ordinary 1-watt flashlight bulb for about 1 second only.

Furthermore, Goldstone engineers estimate that, if Mariner II had continued to function in all its systems and to point its directional antenna at the Earth, useful telemetry data could have been obtained by the DSIF stations out to about 150 to 200 million miles, and tracking data could have been secured from as far as 300 to 400 million miles.

Construction of the DSIF net was begun in 1958. The Goldstone station was ready for the Pioneer III mission in December of that year. In March, 1959, Pioneer IV was successfully tracked beyond the Moon. Later in 1959, Pioneer V was tracked out to over 3 million miles.

Goldstone participated in the 1960 Project Echo communication satellite experiments and the entire net was used in the Ranger lunar missions of 1961-1962. The Goldstone station performed Venus radar experiments in 1961 and 1962 to determine the astronomical unit more precisely and to study the rotation rate and surface characteristics of the planet.

Following the launch of Mariner II on August 27, 1962, the full DSIF net provided 24-hour-per-day tracking coverage throughout the mission except for a few days during the cruise phase. The net remained on the full-coverage schedule through the period of Venus encounter on December 14.

THE GOLDSTONE COMPLEX

The tracking antennas clustered in a 7-mile radius near Goldstone Dry Lake, California, are the central complex of the DSIF net. Three tracking sites are included in the Goldstone Station: Pioneer Site (DSIF 2), Echo Site (DSIF 3), and Venus Site. The Venus Site is used for advanced radar astronomy, communication research experiments, and radio development; it took no direct part in the Mariner spacecraft tracking operations, but was used for the Venus radar experiments.

Pioneer Site has an 85-foot-diameter parabolic reflector antenna and the necessary radio tracking, receiving, and data recording equipment. The antenna can be pointed to within better than 0.02 of a degree. The antenna has one (hour-angle) axis parallel to the polar axis of the Earth, and the other (declination) axis perpendicular to the polar axis and parallel to the equatorial plane of the Earth. This "polar-mount" feature permits tracking on only one axis without moving the other.

The antenna weighs about 240 tons but can be rotated easily at a maximum rate of 1 degree per second. The minimum tracking rate or antenna swing (0.250686486 degree per minute) is equal to the rotation rate of the Earth. Two drive motors working simultaneously but at different speeds provide an antibacklash safety factor. The antenna can operate safely in high winds.

The Pioneer antenna has a type of feed system (Cassegrain) that is essentially similar to that used in many large reflector telescopes. A convex cone is mounted at the center of the main dish. A received signal is gathered by the main dish and the cone, reflected to a subreflector on a quadripod, where the energy is concentrated in a narrow beam and reflected back to the feed collector point on the main dish. The Cassegrain feed system lowers the noise picked up by the antenna by reducing interference from the back of the antenna, and permits more convenient location of components.

The receiving system at Pioneer Site is also equipped with a low-noise, extremely sensitive installation combining a parametric amplifier and a maser. The parametric amplifier is a device that is "pumped" or excited by microwave energy in such a way that, when an incoming signal is at its maximum, the effect is such that the "pumped-in" energy augments the original strength of the incoming signal. At the same time, the parametric amplifier reduces the receiving system's own electronic noise to such a point that the spacecraft can be tracked twice as far as before.

The maser uses a synthetic ruby mixed with chromium and is maintained at the temperature of liquid helium—about 4.7 degrees K or -450 degrees F (just above absolute zero)—and when "pumped" with a microwave field, the molecular energy levels of the maser material are redistributed so as to again improve the signal amplification while lowering the system noise. The maser doubles the tracking capability of the system with a parametric amplifier, and quadruples the capability of the receiver alone.

The antenna output at Pioneer is a wide-band telemetering channel. In addition, the antenna can be aimed automatically, using its own "error signals." At both the Pioneer and Echo sites at Goldstone, however, the antenna is pointed by a punched tape prepared by a special-purpose computer at JPL and transmitted to Goldstone by teletype.

Pioneer Site has a highly sensitive receiver designed to receive a continuous wave signal in a narrow frequency band in the 960-megacycle range. The site has equipment for recording tracking data for use by computers in determining accurate spacecraft position and velocity.

The instrumentation equipment also includes electronic signal processing devices, magnetic-tape recorders, oscillographs, and other supplementary receiving equipment. The telemetered data can be decommutated (recovered from a signal shared by several measurements on a time basis), encoded, and transmitted by teletype in real time (as received from the spacecraft) to JPL.

Echo Site is the primary installation in the Goldstone complex and has antenna and instrumentation facilities identical to those at Pioneer, except that there is no maser amplifier and a simpler feed system is used instead of the Cassegrain. However, Echo was used as a transmitting facility and only as a stand-by receiving station during the Mariner mission.

Echo has a 10-kilowatt, 890-megacycle transmitter which was utilized for sending commands to the Mariner spacecraft. In addition, the site has an "atomic clock" frequency standard, based on the atomic vibrations of rhubidium, which permits high-precision measurements of the radial velocity of the spacecraft. A unit in the Echo system provides for "readback" and "confirmation" by the spacecraft of commands transmitted to it. In a sense, the spacecraft acknowledges receipt of the commands before executing them.

Walter E. Larkin manages the Goldstone Station for JPL.

THE WOOMERA STATION

The Woomera, Australia, Station (DSIF 4), managed by William Mettyear for the Australian Department of Supply, has essentially the same antenna and tracking capabilities as Goldstone Echo Site, but it has no provisions for commanding the spacecraft. A small transmitter is used for tracking purposes only. The station is staffed and operated by the Australian Department of Supply.

The Mobile Tracking Station (DSIF 1) follows the fast-moving spacecraft during its first low-altitude pass over South Africa.

Station 5 of the DSIF is located near Johannesburg in South Africa.

DSIF 4, at Woomera, dominates the landscape in Australia's "outback."

Woomera, like Johannesburg, is capable of receiving tracking (position and velocity) data and telemetered information for real-time transmission by radio teletype to JPL.

THE JOHANNESBURG STATION

DSIF 5 is located just outside Johannesburg in the Republic of South Africa. This station is staffed by the National Institute of Telecommunications Research (NITR) of the South African Council for Scientific and Industrial Research and managed by Douglas Hogg.

The antenna and receiving equipment are identical to the Goldstone Echo Site installation except for minor details. The station has both transmitting and receiving capability and can send commands to the spacecraft. Recorded tracking and telemetered data are transmitted in real time to JPL by radio teletype.

MOBILE TRACKING STATION

The Mobile Tracking Station (DSIF 1) is a movable installation designed for emplacement near the point of injection of a space probe to assist the permanent stations in early acquisition of the spacecraft. This station is necessary because at this point the spacecraft is relatively low in altitude and consequently appears to move very fast across the sky. The Mobile Tracking Station has a fast-tracking antenna for use under these conditions. DSIF 1 was located near the South African station for Mariner II. It has a 10-foot parabolic antenna capable of tracking at a 10-degree-per-second rate. A 25-watt, 890-megacycle transmitter is used for obtaining tracking information. A diplexer permits simultaneous transmission and reception on the same antenna without interference.

The equipment is installed in mobile vans so that the station can be operated in remote areas. The antenna is enclosed in a plastic dome and is mounted on a modified radar pedestal. The radome is inflatable with air and protects the antenna from wind and weather conditions.

These stations of the DSIF tracked Mariner II in flight and sent commands to the spacecraft for the execution of maneuvers. The telemetry data received from the spacecraft during the 129 days of its mission were recorded and transmitted to JPL, where the information was processed and reduced by the computers of the space flight operations complex.

CHAPTER 7
THIRTEEN MILLION WORDS

The task of receiving, relaying, processing, and interpreting the data coming in simultaneously on a twenty-four-hour basis for several months from the several scientific and many engineering sources of the Mariner spacecraft was of truly monumental proportions.

This activity involved five DSIF tracking stations scattered around the world, a communication network, two computing stations and auxiliary facilities, and some 400 personnel over a four-month period.

Although the Mariner scientific information could be stored and subsequently processed at a later (non-real) time, it was necessary to make tracking and position data available almost as soon as it was received (in real time) so that the midcourse maneuver might be computed and transmitted to the spacecraft, and to further perfect the predicted trajectory and arrival time at Venus.

The engineering performance of the many spacecraft subsystems was also of vital concern. Inaccurate operation in any of several areas could endanger the success of the entire mission. The performance of the attitude control system, the Earth and Sun sensors, the power system, and communications were all of critical importance. Corrective action was possible in certain subsystems where trouble could be predicted from the data or where limited breakdown had occurred.

To integrate all the varied activities necessary to accomplish the mission objectives, an organization was formed within JPL to coordinate the DSIF, the communication network, the work of engineering and scientific advisory panels, and the computer facilities required to evaluate the data.

This organization was known as the Space Flight Operations Complex. For operational purposes only, it included the Space Flight Operations Center, a Communication Center, and a Central Computing Facility (CCF). The DSIF was responsive to the requirements of the organization, but was not an integral part of it.

A space flight operations director was responsible for integrating these many functions into a world-wide Mariner space-flight organization. It was an exhausting 109-day task, one that would severely tax all the resources of JPL in terms of know-how, qualified personnel, time, and equipment before Mariner completed its encounter with Venus.

COMMUNICATION CONTROL

The Communication Center at JPL in Pasadena was one of the most active areas during the many days and nights of the Mariner II mission. All of the teletype and radio lines from the Cape, South Africa, Australia, and Goldstone terminated in this Center. A high-speed data line bypassed the Communication Center, linking Goldstone directly with the Central Computing Facility for quick, real-time computer processing of vital flight information.

From the Communication Center, the teletype data and voice circuits were connected to the several areas within JPL where the mission-control activities were centered, and where the data output was being studied.

The Communication Center was equipped with teletype paper-page printers and paper-tape hole reperforators, which received and transmitted data-word and number groups. The teletype lines terminating at the Center included circuits from Goldstone, South Africa, Australia, and Cape Canaveral.

There were three lines to Goldstone for full-time, one-way data transmission. Duplex (simultaneous two-way) transmission was available to Woomera and South Africa on a full-time basis. In each case, a secondary circuit was provided to the overseas sites for use during critical periods and in case the primary radio-teletype circuits had transmission difficulties. These secondary circuits used different radio transmission paths in order to reduce the chance of complete loss of contact for any extended period of time.

Radio signals from Mariner are received on 85-ft. antenna.

The highly sensitive receiver (shown under test) is located in the control room of the station.

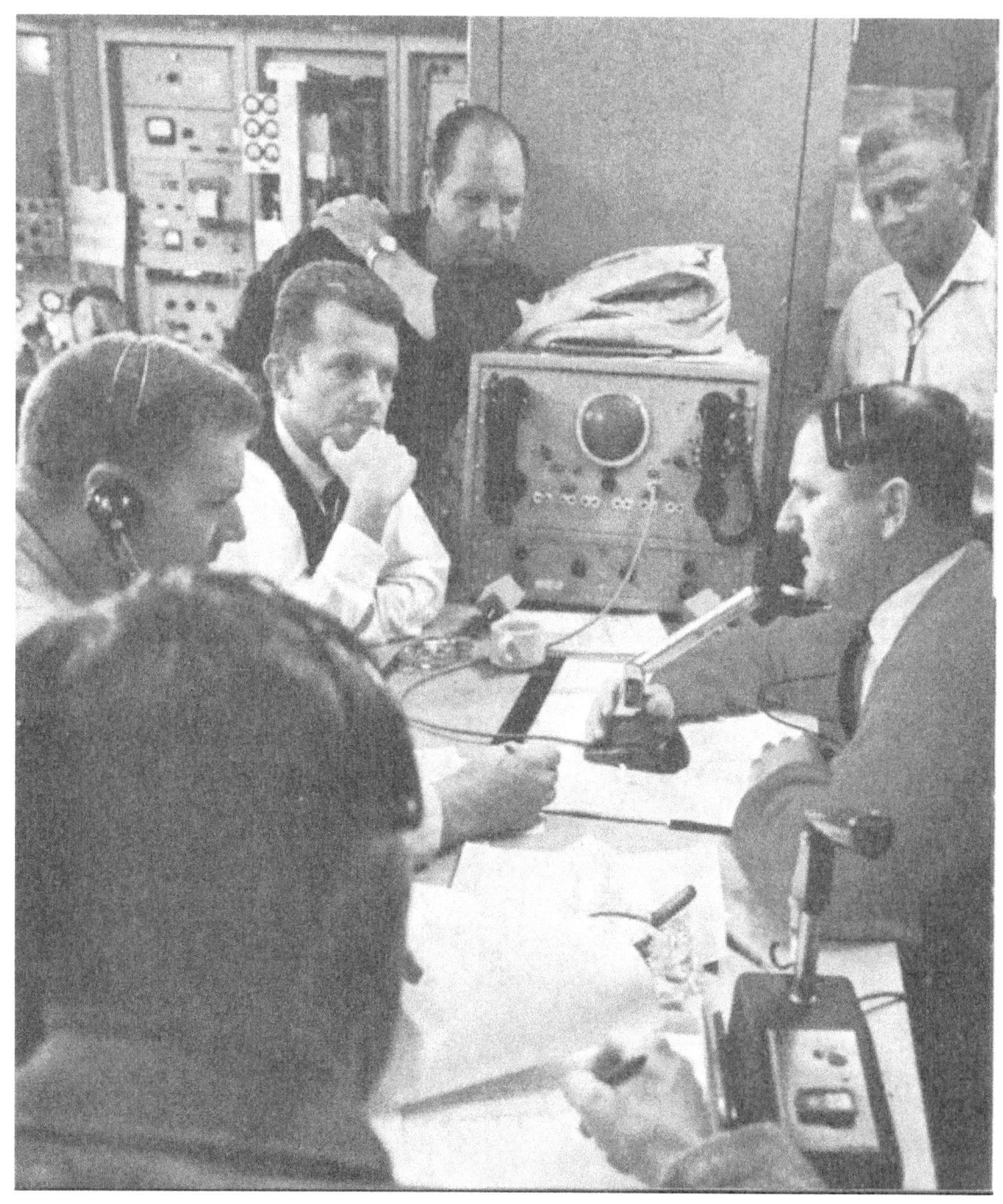

In Goldstone control room, DSIF personnel await confirmation that spacecraft has begun to scan the planet Venus.

```
IZ ZSQSGSGKGKXRGOQOS DNQ XZARZXAVAQQVA XXDRDZ
SQSGIGKGKXRGOQOX DXQ XZAIZXAVQQKDZ XXDRDDDRZ
DXQ XZAIZAZVAQQZDZXXDLKKDRA Z DSQSGIGGGQXRDN
G XXDRZSDRA Z DSQSGIGGGQXRGLZOX DXQ XZAIZAZV
SQSGIGGGQXRGLZOX DXQ XZAIZGAVAQQSXZXXDRDZDRA
DXQ XQAIZGAVQQKVA XXDRDZDRANGDSSQSGIGGGQXRDI
Z XXDRDDDRZLQ ZSQSGIGGGGXRGVZOA DNQ XZARZGAV
QSGIGGGGXRGVQOA DNQ XZARZGAVQKZZAZXXDRZDDRA
DNQ XZARZAZVAKDDGZXXDRDADIZLA DSQSGIGGGGXRGZ
SZXXDRDDDRXOS DSQSQIGGGGXRDDZOA DNQ XZARZXAV
QSGSGGGGXRDSQOG DXQ XZARZXAVAKDZDZXXDRDLDRA
DXQ XZARZXAZZKZZDZXXDLKADRA Z DSQSGIGGGGXRGX
GZXXDRZXDRZ Z DSQSAQGGGGXRDAZOG DXQ XZAIZAZV
DRDXDRA Z DSQSGIGGGGXRDGSOG DXQ XZAIZAZVQKZS
OGGGXRDGAOG DXQ XZAIZAZVQKZVA XXDRDZDRANGDSS
XZAIZXAZZKZZD XXDRDLZDZHA DSQSGIGGGGXRDQAOQ
DLKADRKHQ ZSQSGIGGGGXRDKZOQ DXQ XZAVZAAVQKZZ
GGGGXRS ZOQ DXQ XZAIZAZVQKZDG XXDRDSDIZ A ZS
XZAIZAZVQQKSX XXDRDDDRZNS ZSQSGIGGGGXRSOAOQ
DRDZDRANGDSSQSAQGGGGXRSOAOK DNQ XZAVZAAVQQKZ
GGGGXRSNZOK DNQ XZAIZAZZSQQZSZXXDLKADRA Z DS
XZAIZGAVQQKZG XXDRZXDRZ Z DSQSAQGGGGXRSLAOK
DRDSDRZ Z DSQSAQGGGGXRSLAOK DNQ XZAIZAZZZQKS
GGGGXRSLAOK DNQ XZAIZAZZZQKVA XXDRDZDRANGDSS
XRSIZHZ DXQ XZAIZAZZZQKQQZXXDRDNDRZNQ DSQSGS
ZXAZSQQZXZXXDLKQDRSQS ZSQSGIGGGQXRSVAHZ DXQ
```

From DSIF stations, the data are teletyped in coded format to Pasadena.

Messages are received and routed at the JPL Communications Center.

Data are routed to the digital computer at JPL.

Printout data are made available to experimenters.

Spacecraft status is posted in Operations Center.

The Mobile Tracking Station in South Africa used the Johannesburg communication facilities.

Two one-way circuits for testing and control purposes were open to Cape Canaveral from a month before until after the spacecraft was launched. Lines from the Communication Center to the Space Flight Operations Center at JPL terminated in page printers and reperforators in several locations.

Voice circuits connected all of the stations with Operations Center through the Communication Center. Long-distance radio telephone calls were placed to South Africa to establish contact before the launch sequence was started. Woomera used the Project Mercury voice circuits to the United States during launch and for three days after.

THE OPERATIONS CENTER

The actual nerve center of the Mariner operation was the Space Flight Operations Center (SFOC) at Pasadena. Here, technical and scientific advisory panels reported to the Project Manager and the Mariner Test Director on the performance of the spacecraft in flight, analyzed trajectories, calculated the commands for the midcourse trajectory correction, and studied the scientific aspects of the mission.

These panels were a Spacecraft Data Analysis Team, a Scientific Data Group, an Orbit Determination Group, a Tracking Data Analysis Group, and a Midcourse Command Group.

The Spacecraft Data Analysis Team analyzed the engineering data transmitted from the spacecraft to evaluate the performance of the subsystems in flight. The Team was composed of one or more of the engineers responsible for each of the spacecraft subsystems, and a chairman.

The Science Data Group was composed of the project scientist and certain other scientists associated with the experiments on board the spacecraft. This Group evaluated the data from the scientific experiments while Mariner was in flight and advised the Test Director on the scientific status of the mission.

The Science Data Group was on continuous duty until 48 hours after launch, and at other times during the mission. During encounter with Venus, the Group was also in contact with the scientific experimenters from other participating organizations who were working with JPL.

Closed circuit television monitors are used for instant surveillance of the internal activities of the Operations Center.

A Tracking Data Analysis Group analyzed the tracking data to be used in orbit determination. They also assessed the performance of the DSIF facilities and equipment used to obtain the data.

The Orbit Determination Group used the tracking data to produce estimates of the actual spacecraft trajectory, and to compute the spacecraft path with respect to the Earth, Venus, and the Sun. These calculations were made once each day before the midcourse maneuver, once a week during the cruise phase, and daily during and immediately after the planet encounter.

The Operations Center was equipped with lighted boards on which the progress of the mission was displayed. This information included trajectory data, spacecraft performance, temperature and pressure readings, and other data telemetered from the spacecraft subsystems.

Closed-circuit television was used for coordinating the activities of the SFOC. Operating personnel could use television monitors in four consoles which were linked to six fixed cameras viewing teletype page printers. The entire Operations Room could be kept under surveillance by the Project Manager, the Test Director, or the DSIF Operations Manager, using cameras controlled in "pan," "tilt," and "zoom."

CENTRAL COMPUTING FACILITY

During the Mariner II mission, the JPL Central Computing Facility (CCF) processed approximately 13.1 million data words, or over 90 million binary bits of computer data. (Binary bit = a discrete unit of information intelligible to a digital computer. One data word = 7 binary bits.)

In the four-month operation, about 100,000 tracking and telemetering data cards were received and processed, yielding over 1.2 million computer pages of tabulated, processed, and analyzed data for evaluation by the engineers and scientists. Approximately 1,000 miles of magnetic tape were used in the 1,056 rolls recorded by the DSIF.

The Central Computing Facility processed and reduced tracking and telemetry data from the spacecraft, as recorded and relayed by the stations of the DSIF. The tracking information was the basis for orbital calculations and command decisions. After delivery of telemetry data on magnetic tapes by the DSIF, the CCF stored the data for later reduction and analysis. Where telemetry data were being processed in real or near-real time, certain critical engineering and scientific functions were programmed to print-out an "alarm" reading when selected measurements in the data were outside specified limits.

The CCF consists of three stations at JPL: Station C, the primary computing facility; Station D, the secondary installation; and the Telemetry Processing Station (TPS).

Station C was the principal installation for processing both tracking and telemetry data received from the DSIF tracking stations, both in real and non-real time. The Station was equipped with a high-speed, general-purpose digital computer with a 32,168-word memory and two input-output channels, each able to handle 6 tape units. The associated card-handling equipment was also available.

Tape translators or converters were provided for converting teletype data and other digital information into magnetic tape format for computer input. The teletype-to-tape unit operated at a rate of 300 characters per second.

A smaller computer acted as a satellite of the larger unit, performing bookkeeping and such related functions as card punching, card reading, and listing.

A high-speed unit microfilmed magnetic-tape printout was received from the large computer. It provided "quick-look" copy within 30 minutes of processing the raw data. Various paper-tape-to-card and card-to-paper-tape converters were used to eliminate human error in converting teletype data tape to computer cards.

Station C also utilized another computer as a real-time monitor and to prepare a magnetic tape file of all telemetered measurements for input to the large computer.

Station D was the secondary or backup computational facility, primarily intended for use in case of equipment failure in Station C. During certain critical phases of the Mariner mission—launch, orbit determination, midcourse maneuver—this facility paralleled the operations in Station C.

Station D is equipped with three computers and various card-to-tape converters and teletype equipment.

The Telemetry Processing Station received and processed all demodulated data (that recovered from the radio carrier) on magnetic tapes recorded at the DSIF stations. The TPS output was digital magnetic tapes suitable for computer entry.

The TPS equipment included FM discriminators, a code translator, a device for converting data from analog to digital form, and magnetic-tape recorders. Basically, the equipment accepted the digital outputs from the tape units, the analog-to-digital converter, and the code translator and put them in digital tape format for the computer input.

As the launch operation started on August 27, the powered-flight portion of the space trajectories program was run at launch minus 5 minutes (L minus 5) and was repeated several times because of holds at AMR. The orbit determination program was run at lift-off to calculate the first orbit predictions used for aiding the DSIF in finding the spacecraft in flight.

During the 12 hours following launch, both C and D Stations performed parallel computations on tracking data. Station D discontinued space flight operations at L plus 12 hours and resumed at the beginning of the midcourse maneuver phase.

Tracking data processing and midcourse maneuver studies were conducted daily until the midcourse maneuver was performed at L plus eight days. For the following 97 days, tracking data were processed once each week for orbit determination. Starting three days before the encounter, tracking data were processed daily until the beginning of the encounter phase.

Tracking data processing was conducted in near-real time throughout encounter day, and daily for two days thereafter. For these three days, tracking data were handled in Station D in order to permit exclusive use of Station C for telemetry data processing and analysis. After this three-day period, including the encounter, Station C processed the tracking data every sixth day until the mission terminated on L plus 129 days.

Telemetry data were processed in a different manner. Following the launch, DSIF Station 5 at South Africa received the telemetry signal first, demodulated it, and put it in the proper format for teletype transmission to JPL. The other DSIF stations followed in sequence as the spacecraft was heard in other parts of the world. For two days after launch, the computers processed telemetry data as required by the Spacecraft Data Analysis Team.

During those periods when the large computer was processing tracking data, a secondary unit supplied quick-look data in near-real time. When Goldstone was listening to the spacecraft, quick-look data were processed in real time, using the high-speed data line direct to the Central Computing Facility.

For the 106 days that Mariner was actually in Mode II (cruise), the telemetry data were processed twenty-four hours a day, seven days a week. Data were presented to the engineering and science analysis teams in quick-look format every three hours, except for short maintenance interruptions, one computer failure, and a major modification requiring three days, when a back-up data process mode of operation was used. The large computer performed full processing and analysis of engineering and science data seven days a week from launch until the Venus encounter.

On encounter day, the secondary Station C computer processed telemetry data from the high-speed Goldstone line. Data on magnetic tapes produced by the computer were processed and analyzed by the large unit in near-real time every 30 minutes. The computer processing and delivery time during this operation varied from 4½ to 7 minutes.

CHAPTER 8
THE SCIENTIFIC EXPERIMENTS

After a year of concentrated effort, in which the resources of NASA, the Jet Propulsion Laboratory, and American science and industry had been marshalled, Mariner II had probed secrets of the solar system some billions of years old.

Scientists and engineers had studied the miles of data processed in California from the tapes recorded at the five DSIF tracking stations around the world. Two and a half months of careful analysis and evaluation yielded a revised estimate of Venus and of the phenomena of space. As a result, the dynamics of the solar system were revealed in better perspective and the shrouded planet stood partially unmasked. When the Mariner data were correlated with the data gathered by JPL radar experiments at Goldstone in 1961 and 1962, the relationships between the Earth, Venus, and the Sun became far clearer than ever before.

Two experiments were carried on the spacecraft for a close-up investigation of Venus' atmosphere and temperature characteristics—a microwave radiometer and an infrared radiometer. They were designed to operate during the approximate 35-minute encounter period and at a distance varying from about 10,200 miles to 49,200 miles from the center of the planet. [2]

Cosmic dust detector.

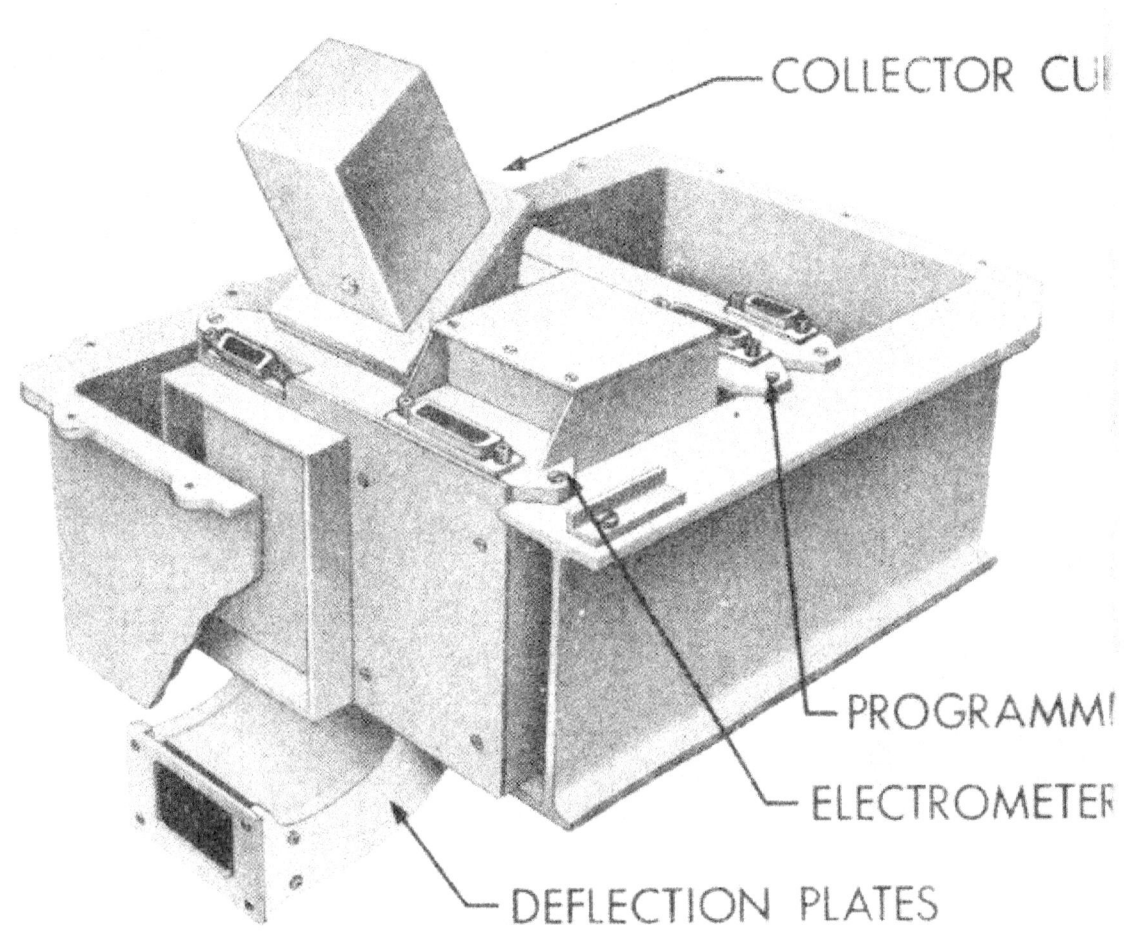

COLLECTOR CU[

PROGRAMM[

ELECTROMETE[

DEFLECTION PLATES

Solar plasma spectrometer.

COLLECTOR CUP

PROGRAMMER

ELECTROMETER

DEFLECTION PLATES

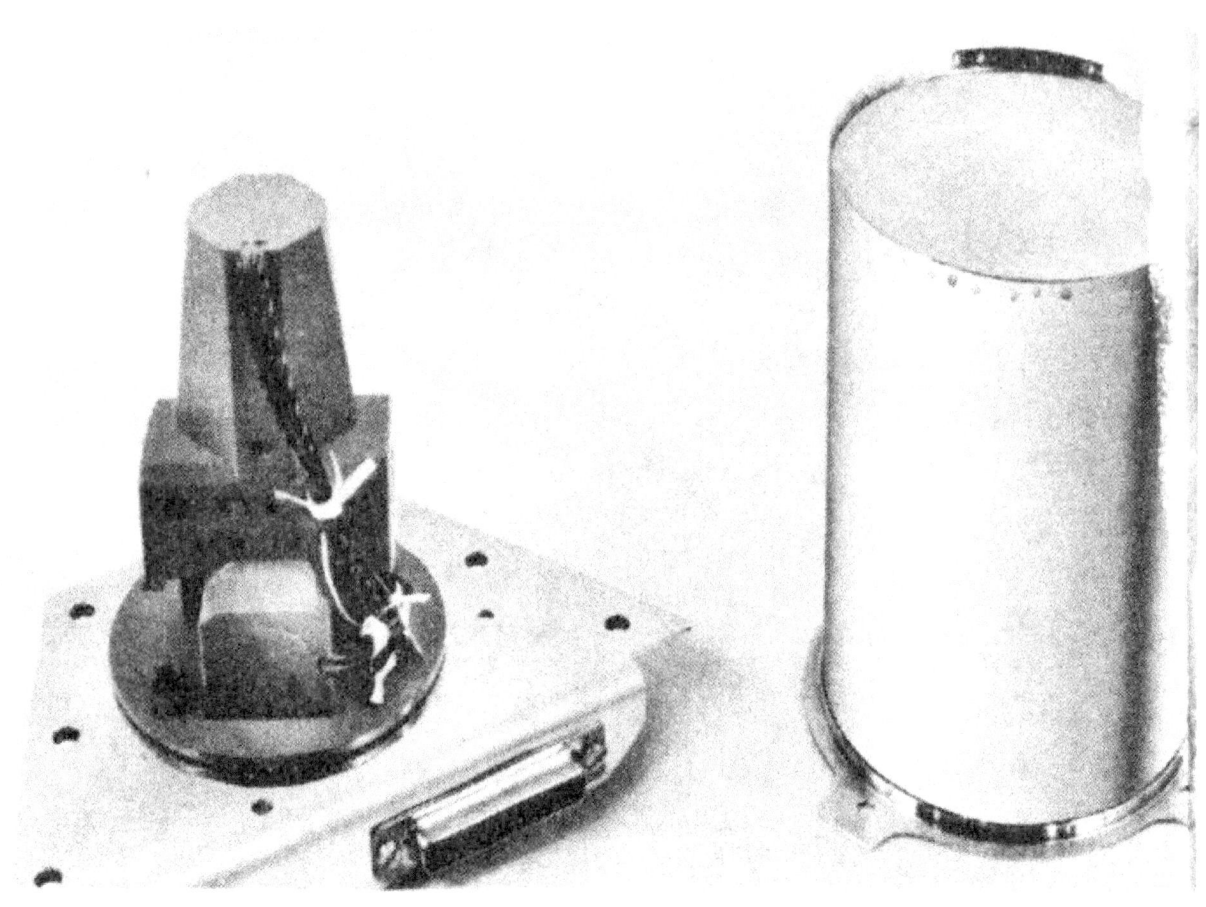

Magnetometer.

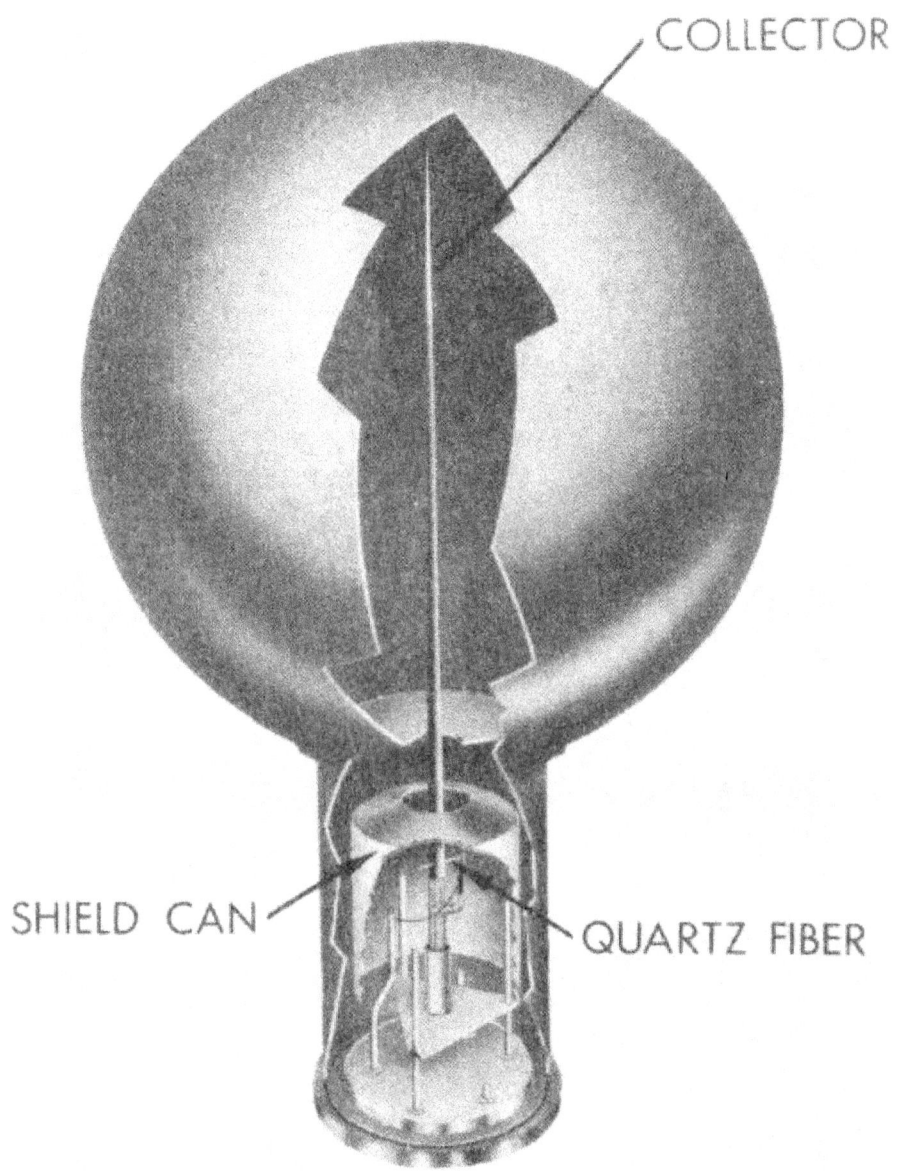

High-energy particle detector.

COLLECTOR

SHIELD CAN

QUARTZ FIBER

Microwave and infrared radiometers.

REFERENCE HORNS

Table 2. Mariner Experiments

Experiment	Description	Experimenters
Microwave radiometer	Determine the temperature of the planet surface and details concerning its atmosphere	Dr. A. H. Barrett, Massachusetts Institute of Technology; D. E. Jones, JPL; Dr. J. Copeland, Army Ordnance Missile Command and Ewen-Knight Corp.; Dr. A. E. Lilley, Harvard College Observatory
Infrared radiometer	Determine the structure of the cloud layer and temperature distributions at cloud altitudes	Dr. L. D. Kaplan, JPL and University of Nevada; Dr. G. Neugebauer, JPL; Dr. C. Sagan, University of California, Berkeley, and Harvard College Observatory
Magnetometer	Measure planetary and interplanetary magnetic fields	P. J. Coleman, NASA; Dr. L. Davis, Caltech; Dr. E. J. Smith, JPL; Dr. C. P. Sonett, NASA
Ion chamber and matched Geiger-Mueller tubes	Measure high-energy cosmic radiation	Dr. H. R. Anderson, JPL; Dr. H. V. Neher, Caltech
Anton special-purpose tube	Measure lower radiation (especially near Venus)	Dr. J. Van Allen and L. Frank, State University of Iowa
Cosmic dust detector	Measure the flux of cosmic dust	W. M. Alexander, Goddard Space Flight Center, NASA
Solar plasma spectrometer	Measure the intensity of low-energy positively charged particles from the Sun	M. Neugebauer and Dr. C. W. Snyder, JPL

Four experiments for investigation of interplanetary space and the regions near Venus employed: a magnetometer; high-energy charged particle detectors, including an ionization chamber and Geiger-Mueller radiation counters; a cosmic dust detector; and a solar plasma detector.

These six scientific experiments represented the cooperative efforts of scientists at nine institutions: The Army Ordnance Missile Command, the Ewen-Knight Corp., the California Institute of Technology, the Goddard Space Flight Center of NASA, Harvard College Observatory, the Jet Propulsion Laboratory, the Massachusetts Institute of Technology, the State Universities of Iowa and Nevada, and the University of California at Berkeley. Table 2 lists the experiments, the experimenters, and their affiliations.

At the Jet Propulsion Laboratory, the integration of the scientific experiments and the generation of a number of them were carried out under the direction of Dr. Manfred Eimer. R. C. Wyckoff was the project scientist and J. S. Martin was responsible for the engineering of the scientific experiments.

DATA CONDITIONING SYSTEM

Mariner's scientific experiments were controlled and their outputs processed by a data conditioning system which gathered the information from the instruments and prepared it for transmission to the Earth by telemetry. In this function, the data system acted as a buffer between the science systems and the spacecraft data encoder.

The pulse output of certain of the science instruments was counted and the voltage amplitude representations of other instruments were converted from analog form to a binary digital equivalent of the information signals. The data conditioning system also included circuits to permit time-sharing of the telemetry channels with the spacecraft engineering data, generation of periodic calibration signals for the radiometer and magnetometer, and control of the direction and speed of the radiometer scanning cycle.

During Mariner's cruise mode, the data conditioning system was used for processing both engineering and science data. If the spacecraft lost lock on the Sun or the Earth during the cruise mode, no scientific data would be telemetered during the reorientation period. Engineering data were sampled and transmitted for about 17 seconds during every 37-second interval. The planetary encounter mode involved only science and no engineering data transmission. In this mode, the science data were sampled during 20-second intervals.

COSMIC DUST DETECTOR

The cosmic dust detector on Mariner II was designed to measure the flux density, direction, and momentum of interplanetary dust particles between the Earth and Venus. These measurements were concerned with the particles' direction and distance from the Sun, the momentum with respect to the spacecraft, the nature of any concentrations of the dust in streams, variations in cosmic dust flux with distance from the Earth and Venus, and the possible effects on manned flight.

Mariner's cosmic dust instrument could detect a particle as small as something like a billionth of a gram, or about five-trillionths of a pound. This type of sensor had been used on rockets even before Explorer I. It had yielded good results on Pioneer I in the region between the Earth and the Moon. The instrument was a 55-square-inch acoustical detector plate, or sounding board, made of magnesium. A crystal microphone was attached to the center of the plate. The instrument could detect both low- and high-momentum particles and also provide a rough idea of their direction of travel.

The dust particle counters were read once each 37 seconds during the cruise mode. This rate was increased to once each 20 seconds during the encounter with Venus.

The instrument was attached to the top of the basic hexagonal structure; it weighed 1.85 pounds, and consumed only 0.8 watt of power.

SOLAR PLASMA EXPERIMENT

In order to investigate the phenomena associated with the movement of plasma (charged particles of low energy and density streaming out from the Sun to form the so-called "solar wind") in interplanetary space, Mariner carried a solar plasma spectrometer that measured the flux and energy spectrum of positively charged plasma components with energies in the range 240 to 8400 volts. The extremely sensitive plasma detector unit was open to space, consumed 1 watt of power, and consisted of four basic elements: curved electrostatic deflection plates and collector cup, electrometer, a sweep amplifier, and a programmer.

The curved deflector plates formed a tunnel that projected from the chassis on the spacecraft hexagon in which the instrument was housed. Pointed toward the Sun, the gold-plated magnesium deflector plates gathered particles from space. Since the walls of the tunnel each carried different electrical charges, only particles with just the correct energy and speed could pass through and be detected by the collector cup without striking the charged walls. A sensitive electrometer circuit then measured the current generated by the flow of the charged particles reaching the cup.

The deflection plates were supplied by amplifier-generated voltages which were varied in 10 steps, each lasting about 18 seconds, allowing the instrument to measure protons with energies in the 240 to 8,400 electron volt range. The programmer switched in the proper voltage and resistances.

HIGH-ENERGY RADIATION EXPERIMENT

Mariner carried an experiment to measure high-energy radiation in space and near Venus. The charged particles measured by Mariner were primarily cosmic rays (protons or the nuclei of hydrogen atoms), alpha particles (nuclei of helium atoms), the nuclei of other heavier atoms, and electrons. The study of these particles in space and those which might be trapped near Venus was undertaken in the hope of a better understanding of the dynamics of the solar system and the potential hazards to manned flight.

The high-energy radiation experiment consisted of an ionization chamber and detectors measuring particle flux (velocity times density), all mounted in a box measuring 6 × 6 × 2 inches and weighing just under 3 pounds. The box was attached halfway up the spacecraft superstructure in order to isolate the instruments as much as possible from secondary emission particles produced when the spacecraft was struck by cosmic rays, and to prevent the spacecraft from blocking high-energy radiation from space.

The ionization chamber had a stainless steel shell 5 inches in diameter, with walls only 1/100-inch thick. The chamber was filled with argon gas into which was projected a quartz fibre next to a quartz rod.

A charged particle entering the chamber would leave a wake of ions in the argon gas. Negative ions accumulated on the rod, reducing the potential between the rod and the spherical shell, eventually causing the quartz fibre to touch the rod. This action discharged the rod, producing an electrical pulse which was amplified and transmitted to the Earth. The rod was then recharged and the fibre returned to its original position.

In order to penetrate the walls of the chamber, protons required an energy of 10 million electron volts (Mev), electrons needed 0.5 Mev, and alpha particles 40 Mev.

The particle flux detector incorporated three Geiger-Mueller tubes, two of which formed a companion experiment to the ionization chamber; each generated a current pulse whenever a charged particle was detected. One tube was shielded by an 8/1,000-inch-thick stainless steel sleeve, the other by a 24/1,000-inch-thick electron-stopping beryllium shield. Thus, the proportion of particles could be determined.

The third Geiger-Mueller tube was an end-window Anton-type sensor with a mica window that admitted protons with energies greater than 0.5 Mev and electrons, 40,000 electron volts. A magnesium shield around the rest of the tube enabled the instrument to determine the direction of particles penetrating only the window.

The three Geiger-Mueller tubes protruded from the box on the superstructure of the spacecraft. The end-window tube was inclined 20 degrees from the others and 70 degrees from the spacecraft-Sun line since it had to be shielded from direct solar exposure.

THE MAGNETOMETER

Mariner carried a magnetometer to measure the magnetic field in interplanetary space and in the vicinity of Venus. Lower sensitivity limit of the instrument was about 5 gamma. A gamma is a unit of magnetic measurement and is equal to 10^{-5} or 1/100,000 oersted, or 1/30,000 of the Earth's magnetic field at the equator. The nails in one of your shoes would probably produce a field of about 1 gamma at a distance of approximately 4 feet.

Housed in a 6- × 3-inch metal cylinder, the instrument consisted of three magnetic core sensors, each aligned on a different axis to read the three magnetic field components and having primary and secondary windings. The presence of a magnetic field altered the current in the secondary winding in proportion to the strength of the field encountered.

The magnetometer was attached near the top of the superstructure, just below the omni-antenna, in order to remove it as far as possible from any spacecraft components having magnetic fields of their own.

An auxiliary coil was wound around each of the instrument's magnetic sensor cores to compensate for permanent magnetic fields existing in the spacecraft itself. These spacecraft fields were measured at the magnetometer before launch and, in flight, the auxiliary coils carried currents of sufficient strength to cancel out the spacecraft's magnetic fields.

The magnetometer reported almost continuously on its journey and for 20 days after encounter. During the encounter, observations were made each 20 seconds on each of the three components of the magnetic field.

MICROWAVE RADIOMETER

A microwave radiometer on board Mariner II was designed to scan Venus during encounter at two wavelengths: 13.5 and 19 millimeters. The radiometer was intended to help settle some of the controversies about the origin of the apparently high surface temperature emanating from Venus, and the value of the surface temperature.

The equipment included a 19-inch-diameter parabolic antenna mounted above the basic hexagonal structure on a swivel driven in a 120-degree scanning motion by a motor. The radiometer electronics circuits were housed behind the antenna dish. The antenna was equipped with a diplexer, which allowed it to receive both wavelengths at once without interference, and to compare the signals emanating from the two reference horns with those from the planet. The reference horns were pointed away from the main antenna beam so they would look into deep space as Mariner passed Venus. This feature allowed the antenna to "bring in" a reference temperature of approximately absolute zero during encounter.

The microwave radiometer was to be turned on 10 hours before the encounter began. An electric motor was then to start a scanning or "nodding" motion of 120 degrees at the rate of 1 degree per second. Upon radiometer contact with the planet, this scanning rate would be reduced to 1/10 degree per second as long as the planetary disk was scanned. A special command system in the data conditioning system would reverse or normalize the direction of scan as the radiometer reached the edge or limb of the planet.

The signals from the antenna and the reference horns were to be processed and the data handled in a receiver, located behind the antenna, which measured the difference between the signals from Venus and the reference signals from space. The information was then to be telemetered to the Earth.

The microwave radiometer was automatically calibrated twenty-three times during the mission by a sequence originating in the data conditioning system, so that the correct functioning of the instrument could be determined before the encounter with Venus.

INFRARED RADIOMETER

The infrared radiometer was a companion experiment to the microwave instrument and was rigidly mounted to the microwave antenna so that both radiometers would look at the same area of Venus with the same scanning rate. The instrument detected radiation in the 8 to 9 and 10 to 10.8 micron regions of the infrared spectrum.

The infrared radiometer had two optical sensors. As the energy entered the system, it was "chopped" by a rotating disk, alternately passing or comparing emissions from Venus and from empty space. The beam was then split by a filter into the two wavelength regions. The output was then detected, processed, and transmitted to the Earth.

The infrared radiometer measured 6 inches by 2 inches, weighed 2.7 pounds, and consumed 2 watts of power. The instrument was equipped with a calibration plate which was mounted on a superstructure truss adjacent to the radiometer.

MARINER'S SCIENTIFIC OBJECTIVES

Equipped with these instruments and with the mechanism for getting the measurements back to Earth, Mariner II was prepared to look for the answers to some of the questions inherent in its over-all mission objectives:

1. The investigation of interplanetary space between the Earth and Venus, measuring such phenomena as the cosmic dust, the mysterious plasma or solar winds, high-energy cosmic rays from space outside our solar system, charged particles from the Sun, and the magnetic fields of space.

2. The experiments to be performed near Venus (at about 21,150 miles out from the surface) in an effort to understand its magnetic fields, radiation belts, the temperature and composition of its clouds, and the temperature and conditions on the surface of the planet.

CHAPTER 9
THE LEGACY OF MARINER

If intelligent life had existed on Venus on the afternoon of the Earth's December 14, 1962, and if it could have seen through the clouds, it might have observed Mariner II approach from the night side, drift down closer, cross over to the daylight face, and move away toward the Sun to the right. The time was the equivalent of 12:34 p.m. along the Pacific Coast of the United States, where the spacecraft was being tracked.

Mariner II had reached the climax of its 180-million-mile, 109-day trip through space. The 35-minute encounter with Venus would tell Earth scientists more about our sister planet than they had been able to learn during all the preceding centuries.

SPACE WITHOUT DUST?

Before Mariner, scientists theorized about the existence of clouds of cosmic dust around the Sun. A knowledge of the composition, origin, and the dynamics of these minute particles is necessary for study of the origins and evolution of the solar system.

Tiny particles of cosmic dust (some with masses as low as 1.3×10^{-10} gram or about one-trillionth of a pound) were thought to be present in the solar system and have been recorded by satellites in the near-Earth regions.

These microcosmic particles could be either the residue left over after our solar system was formed some 5 billion years ago, possibly by condensation of huge masses of gas and dust clouds; or, the debris deposited within our system by the far-flung and decaying tails of passing comets; or, the dust trapped from galactic space by the magnetic fields of the Sun and the planets.

Analysis of the more than 1,700 hours of cosmic dust detector data recovered from the flight of Mariner II seems to indicate that in the region between the Earth and Venus the concentration of tiny cosmic dust particles is some ten-thousand times less than that observed near the Earth.

During the 129 days (including the post-encounter period) of Mariner's mission, the data showed only one dust particle impact which occurred in deep space and not near Venus. Equivalent experiments near Earth (on board Earth satellites) have yielded over 3,700 such impacts within periods of approximately 500 hours. The cause of this heavy near-Earth concentration, the exact types of particles, and their source are still unknown.

The cosmic dust experiment performed well during the Mariner mission. Although some calibration difficulty was observed about two weeks before the Venus encounter, possibly caused by overheating of the sensor crystal, there was no apparent effect in the electronic circuits.

THE UBIQUITOUS SOLAR WIND

For some time prior to Mariner, scientists postulated the existence of a so-called plasma flow or "solar wind" streaming out from the Sun, to explain the motion of comet tails (which always point away from the Sun, perhaps repelled by the plasma), geomagnetic storms, aurorae, and other such disturbances. (Plasma is defined as a gas in which the atoms are dissociated into atomic nuclei and electrons, but which, as a whole, is electrically neutral.)

The solar wind was thought to drastically alter the configuration of the Sun's external magnetic field. Plasma moving at extreme velocities is able to carry with it the lines of magnetic force originating in the Sun's corona and to distort any fields it encounters as it moves out from the Sun.

It was believed that these moving plasma currents are also capable of altering the size of a planet's field of magnetic flux. When this happens, the field on the sunlit face of the planet is compressed and the dark side has an elongated expansion of the field. For example, the outer boundary of the Earth's magnetic field is pushed in by the solar wind to about 40,000 miles from the Earth on the sunward side. On the dark side, the field extends out much farther.

The solar wind was also known to have an apparent effect on the movement of cosmic rays. As the Sun spots increase in the regular 11-year cycle, the number of cosmic rays reaching the Earth from outside our solar system will decrease.

Mariner II found that streams of plasma are constantly flowing out from the Sun. This fluctuating, extremely tenuous solar wind seems to dominate interplanetary space in our region of the solar system. The wind moves at velocities varying from about 200 to 500 miles per second (about 720,000 to 1,800,000 miles per hour), and measures up to perhaps a million degrees Fahrenheit (within the subatomic structure).

With the solar plasma spectrometer working at ten different energy levels, Mariner required 3.7 minutes to run through a complete energy spectrum. During the 123 days, when readings were made, a total of 40,000 such spectra were recorded. Plasma was monitored on 104 of those 123 days, and on every one of the spectra, the plasma was always present.

Mariner showed that the energies of the particles in the solar winds are very low, on the order of a few hundred or few thousand electron volts, as compared with the billions and trillions of electron volts measured in cosmic radiation.

The extreme tenuousity or low density of the solar wind is difficult to comprehend: about 10 to 20 protons (hydrogen nuclei) and electrons per cubic inch. But despite the low energy and density, solar wind particles in our region of the solar system are billions of times more numerous than cosmic rays and, therefore, the total energy content of the winds is much greater than that of the cosmic rays.

Mariner found that when the surface of the Sun was relatively inactive, the velocity of the wind was a little less than 250 miles per second and the temperature a few hundred thousand degrees. The plasma was always present, but the density and the velocity varied. Flare activity on the Sun seemed to eject clouds of plasma, greatly increasing the velocity and density of the winds. Where the particles were protons, their energies ranged from 750 to 2,500 electron volts.

The experiment also showed that the velocity of the plasma apparently undergoes frequent fluctuations of this type. On approximately twenty occasions, the velocity increased within a day or two by 20 to 100%. These disturbances seemed to correlate well with magnetic storms observed on the Earth. In several cases, the sudden increase in the solar plasma flux preceded various geomagnetic effects observed on the Earth by only a short time.

The Mariner solar plasma experiment was the first extensive measurement of the intensity and velocity spectrum of solar plasma taken far enough from the Earth's field so that the Earth would have no effect on the results.

HIGH-ENERGY PARTICLES: FATAL DOSAGE?

Speculation has long existed as to the amount of high-energy radiation (from cosmic rays and particles from the Sun with energies in the millions of electron volts) present within our solar system and as to whether exposure would be fatal to a human space traveler.

This high-energy type of ionizing radiation is thought to consist of the nuclei of such atoms as hydrogen and helium, and of electrons, all moving very rapidly. The individual particles are energetic enough to penetrate considerable amounts of matter. The concentration of these particles is apparently much lower than that of low-energy plasma.

The experiments were designed to detect three types of high-energy radiation particles: the cosmic rays coming from outside the solar system, solar flare particles, and radiation trapped around Venus (as that which is found in the Earth's Van Allen Belt).

These high-energy radiation particles (also thought to affect aurorae and radio blackouts on the Earth) measure from about one hundred thousand electron volts up to billions of volts. The distribution of this energy is thought to be uniform outside the solar system and is assumed to move in all directions in a pattern remaining essentially constant over thousands of years.

Inside the solar system, the amount of such radiation reaching the Earth is apparently controlled by the magnetic fields found in interplanetary space and near the Earth.

The number of cosmic rays changes by a large amount over the course of an 11-year Sun-spot cycle, and below a certain energy level (5,000 Mev) few cosmic rays are present in the solar system. They are probably deflected by plasma currents or magnetic fields.

Mariner's charged particles experiment indicated that cosmic radiation (bombardment by cosmic rays), both from galactic space and those particles originating in the Sun, would not have been fatal to an astronaut, at least during the four-month period of Mariner's mission.

The accumulated radiation inside the counters was only 3 roentgens, and during the one solar storm recorded on October 23 and 24, the dosage measured only about ¼ roentgen. In other words, the dosage amounts to about one-thousandth of the usually accepted "half-lethal" dosage, or that level at which half of the persons exposed would die. An astronaut might accept many times the dosage detected by Mariner II without serious effects.

The experiment also showed little variation in density of charged particles during the trip, even with a 30% decrease in distance from the Sun, and no apparent increase due to magnetically trapped particles or radiation belts near Venus as compared with interplanetary space. However, these measurements were made during a period when the Sun was slowly decreasing in activity at the end of an 11-year cycle. The Sun spots will be at a minimum in 1964-1965, when galactic cosmic rays will sharply increase. Further experiments are needed to sample the charged particles in space under all conditions.

The lack of change measured by the ionization chamber during the mission was significant; the cosmic-ray flux of approximately 3 particles per square centimeter per second throughout the flight was an unusually constant value. A clear increase in high-energy particles (10 Mev to about 800 Mev) emitted by the Sun was noted only once: a flare-up between 7:42 and 8:45 a.m., PST, October 23. The ionization chamber reading began to increase before the flare disappeared. From a background reading of 670 ion pairs per cubic centimeter per second per standard atmosphere, it went to a peak value of 18,000, varied a bit, and remained above 10,000 for 6 hours before gradually decreasing over a period of several days. Meanwhile, the flux of the particles detected by the Geiger counter rose from the background count of 3 to a peak of 16 per square centimeter per second. Ionization thus increased much more than the number of particles, indicating to the scientists that the high-energy particles coming from the Sun might have had much lower average energies than the galactic cosmic rays.

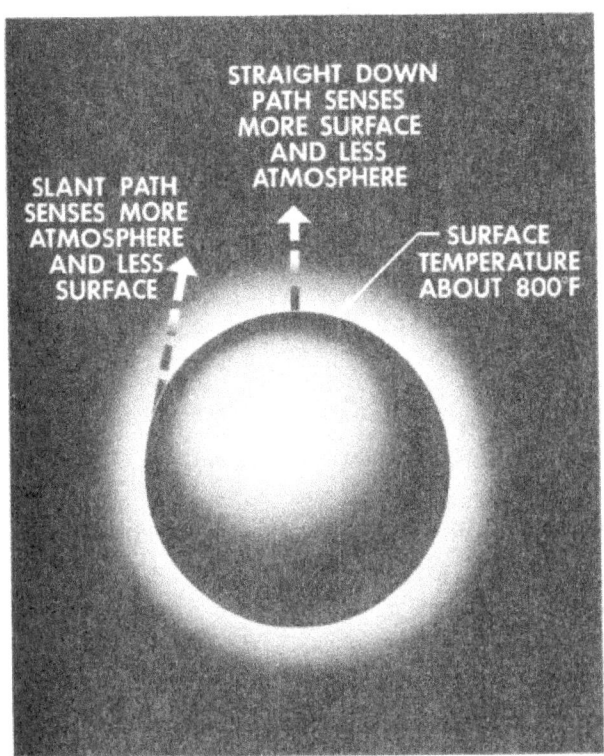
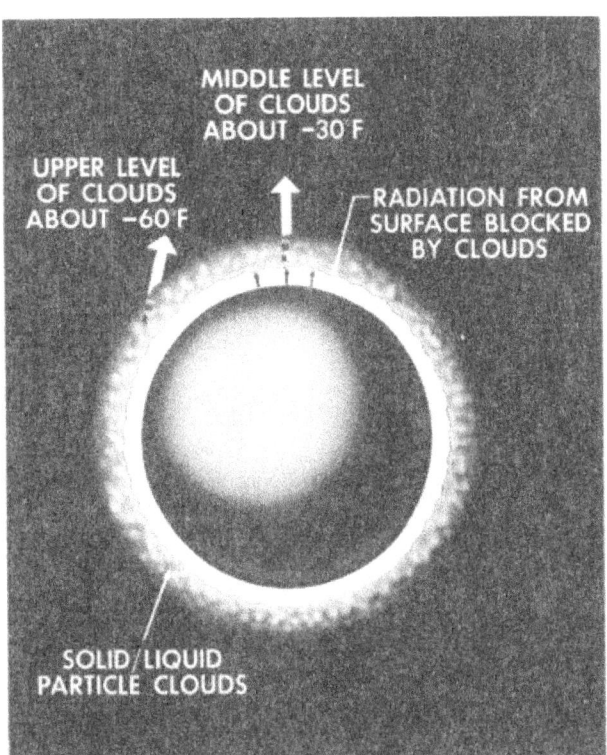

Data obtained by microwave radiometer are illustrated at left; results of infrared radiometer experiment are shown at right. Note how moving spacecraft sees more of atmosphere along limb or edge of planet, less in center.

In contrast, the low-energy experiment detected the October 23 event, and eight or ten others not seen by the high-energy detectors. These must have been low-penetrating particles excluded by the thicker walls of the high-energy instrument. These particles were perhaps protons between 0.5 and 10 Mev or electrons between 0.04 and 0.5 Mev.

At 20,000 miles from the Earth, the rate at which high-energy particles have been observed has been recorded at several thousand per second. With Mariner at approximately the same distance from Venus, the average was only one particle per second, as it had been during most of the month of November in space. Such a rate would indicate a low planetary magnetic field, or one that did not extend out as far as Mariner's 21,598-mile closest approach to the surface.

Mariner II measured and transmitted data in unprecedented quantity and quality during the long trip. In summary, Mariner showed that, during the measuring period, particles were numerous in the energy ranges from a few hundred to 1,000 electron volts. Protons in the range 0.5 to 10 Mev were not numerous, but at times the flux (density) was several times that of cosmic rays.

Almost no protons were shown in the 10 to 800 Mev range, except during solar flares when the particles in this range were numerous. Above 800 Mev (primarily those cosmic rays entering interplanetary space from outside the solar system) the number decreased rapidly as the energy increased, the average total being about 3 per centimeter per second.

During one 30-day period in November and December, the low-energy counter saw only two small increases in radiation intensity. At this time, the mean velocity of the solar wind was considerably lower than during September and October. This might suggest that high-velocity plasma and low-energy cosmic rays might both originate from the same solar source.

A MAGNETIC FIELD?

Prior to the Mariner II mission, no conclusive evidence had ever been presented concerning a Venusian magnetic field and nothing was known about possible fluid motions in a molten core or other hypotheses concerning the interior of the planet.

Scientists assumed that Venus had a field somewhat similar to the Earth's, although possibly reduced in magnitude because of the apparently slow rate of rotation and the pressure of solar plasma. Many questions had also been raised concerning the nature of the atmosphere, charged particles in the vicinity of the planet, magnetic storms, and aurorae. Good magnetometer data from Mariner II would help solve some of these problems.

Mariner's magnetometer experiment also sought verification of the existence and nature of a steady magnetic field in interplanetary space. This would be important in understanding the charged particle balance of the inner solar system. Other objectives of the experiment were to establish both the direction and the magnitude of long-period fluctuations in the interplanetary magnetic field and to study solar disturbances and such problems in magnetohydrodynamics (the study of the motion of charged particles and their surrounding magnetic fields) as the existence and effect of magnetized and charged plasmas in space.

The strength of a planet's field is thought to be closely related to its rate of rotation—the slower the rotation, the weaker the field. As a consequence, if Venus' field is simple in structure like the Earth's, the surface field should be 5 to 10% that of the Earth. If the structure of the field is complex, the surface field in places might exceed the Earth's without increasing the field along Mariner's trajectory to observable values.

Most of the phenomena associated with the Earth's magnetic field are likely to be significantly modified or completely absent in and around Venus. Auroral displays and the trapping of charged particles in radiation belts such as our Van Allen would be missing. The field of the Earth keeps low- and moderate-energy cosmic rays away from the top of the atmosphere, except in the polar regions. The cosmic ray flux at the top of Venus' atmosphere is likely to correspond everywhere to the high level found at the Earth's poles.

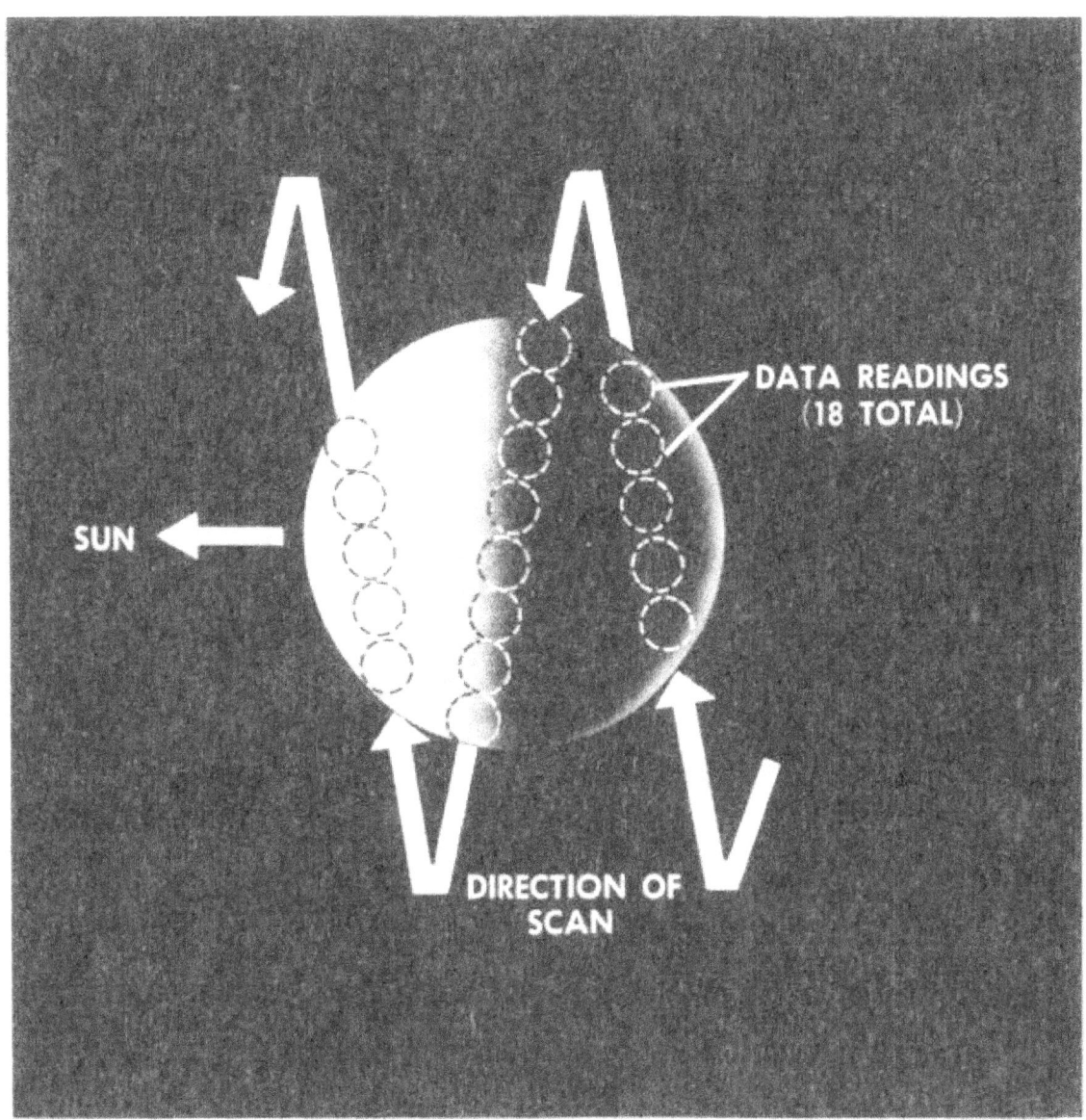

As it encountered Venus, Mariner II made three scans of the planet.

SUN

DIRECTION OF SCAN

DATA READINGS (18 TOTAL)

In contrast to Venus, Jupiter, which is ten times larger in mass and volume and rotates twice as fast as the Earth, has a field considerably stronger than the Earth's. The Moon has a field on the sunlit side (according to Russian measurements) which, because of the Moon's slow rotation rate, is less than ⅓ of 1% of the Earth's at the Equator. Thus, a planet's rotation, if at a less rapid rate than the Earth's, seems to produce smaller magnetic fields.

This theory is consistent with the idea of a planetary magnetic field resulting from the dynamo action inside the molten core of a rotating planet.

The Sun, on the whole, has a fairly regular dipole field. Superimposed on this are some very large fields associated with disturbed regions such as spots or flares, which produce fields of very great intensities.

These solar fields are drawn out into space by plasma flow. Although relatively small in magnitude, these fields are an important influence on the propagation of particles. And the areas in question are very large—something on the order of an astronomical unit.

Mariner II seemed to show that, in space, a generally quiet magnetic-field condition was found to exist, measuring something less than 10 gamma and fluctuating over periods of 1 second to 1 minute.

As Mariner made its closest approach to Venus, the magnetometer saw no significant change, a condition also noted by the radiation and solar plasma detectors. The magnetic field data looked essentially as they had in interplanetary space, without either fluctuations or smooth changes.

The encounter produced no slow changes, nor was there a continuous fluctuation as in the interplanetary regions. There was no indication of trapped particles or near-Venus modification in the flow of solar plasma.

On the Earth's sunny side, a definite magnetic field exists out to 40,000 miles, and on the side away from the Sun considerably farther. If Venus' field had been similar to the Earth's, a reading of 100 to 200 gamma, a large cosmic-ray count, and an absence of solar plasma should have been shown, but none of these phenomena were noted by Mariner.

These results do not prove that Venus definitely has no magnetic field, but only that it was not measurable at Mariner's 21,598-mile point of closest approach. The slow rotation rate and the pressure of the solar winds probably combine to limit the field to a value one tenth of the Earth's. Since Mariner passed Venus on the sunlit side, readings are required on the dark side in order to confirm the condition of the magnetic field on that side of the planet, which normally should be considerably extended.

THE SURFACE: HOW HOT?

Before Mariner, scientists had offered two main theories about the surface of Venus: It had either an electrically charged ionosphere causing false high-temperature readings on Earth instruments despite a cool surface, or a hot surface with clouds becoming increasingly colder with altitude.

The cool-surface theory supposed an ionosphere with a layer of electrons having a density thousands of times that of the Earth's upper atmosphere. Microwave radiations from this electrical layer would cause misleading readings on Earth instruments. As a space probe scanned across such an atmosphere, it would see the least amount of charged ionosphere when looking straight down, and the most concentrated amount while scanning the limb or edge. In the latter case, it would be at an angle and would show essentially a thickening effect of the atmosphere because of the curvature of the planet.

As the probe approached the edge, the phenomenon known as "limb brightening" would occur, since the instruments would see more of the electron-charged ionosphere and little if any of the cooler surface. The temperature readings would, therefore, be correspondingly higher at the limbs.

The other theory, held by most scientists, visualized a hot surface on Venus, with no heavy concentration of electrons in the atmosphere, but with cooler clouds at higher altitudes. Thus, the spacecraft would look at a very hot planet from space, covered by colder, thick clouds. Straight down, the microwave radiometer would see the hot surface through the clouds. When approaching the limb, the radiations would encounter a thicker concentration of atmosphere and might not see any of the hot surface. This condition, "limb darkening," would be characterized by temperatures decreasing as the edges of the planet were approached.

An instrument capability or resolution much higher than that available from the Earth was required to resolve the limb-brightening or limb-darkening controversy. Mariner's radiometer would be able to provide something like one hundred times better resolution than the Earth-based measurements.

At 11:59 a.m., PST, on December 14, 1962, Mariner's radiometers began to scan the planet Venus in a nodding motion at a rate of 0.1 degree per second and reaching an angular sweep of nominally 120 degrees. The radiometers had been switched on 6½ hours before the encounter with Venus and they continued to operate for another hour afterward.

The microwave radiometer looked at Venus at a wavelength of 13.5 millimeters and 19 millimeters. The 13.5-millimeter region was the location of a microwave water absorption band within the electromagnetic spectrum, but it was not anticipated that it would detect any water vapor on Venus. These measurements would allow determination of atmospheric radiation, averaging the hot temperatures near the surface, the warmer clouds at lower levels, and the lower temperatures found in the high atmosphere. If the atmosphere were a strong absorber of microwave energy at 13.5 millimeters, only the temperature of the upper layers would be reported.

Unaffected by water vapor, 19-millimeter radiations could be detected from deeper down into the cloud cover, perhaps from near or at the planet's surface. Large temperature differences between the 19- and 13.5 millimeter readings would indicate the relative amount of water vapor present in the atmosphere. The 19-millimeter radiations would also test the limb-brightening theory.

During its scanning operation, Mariner telemetered back to Earth about 18 digital data points, represented as voltage fluctuations in relation to time. The first scan was on the dark side, going up on the planet: the distance from the surface was 16,479 miles at midscan, and the brightness temperature was 369 degrees F. The second scan nearly paralleled the terminator (junction of light and dark sides) but crossed it going down; it was made from 14,957 miles at midscan and showed a temperature of 566 degrees F. The final scan, 13,776 miles at midpoint, showed 261 degrees F as it swept across the sunlit side of Venus in an upward direction.

The brightness temperature recorded by Mariner's radiometer is not the true temperature of the surface. It is derived from the amount of light or radio energy reflected or emitted by an object. If the object is not a perfect light emitter, as most are not, then the light and radio energy will be some fraction of that returned from a 100% efficient body, and the object is really hotter than the brightness measurement shows. Thus, the brightness temperature is a minimum reading and in this case, was lower than the actual surface temperature.

Mariner's microwave radiometer showed no significant difference between the light and dark sides of Venus and, importantly, higher temperatures along the terminator or night-and-day line of the planet. These results would indicate no ionosphere supercharged with electrons, but a definite limb-darkening effect, since the edges were cooler than the center of the planet.

Therefore, considering the absorption characteristics of the atmosphere and the emissivity factor derived from earlier JPL radar experiments, a fairly uniform 800 degrees F was estimated as a preliminary temperature figure for the entire surface.

Venus is, indeed, a very hot planet.

CLOUD TEMPERATURES: THE INFRARED READINGS

Mariner II took a close look at Venus' clouds with its infrared radiometer during its 35-minute encounter with the planet. This instrument was firmly attached to the microwave radiometer so the two devices would scan the same areas of Venus at the same rate and the data would be closely correlated. This arrangement was necessary to produce in effect a stereoscopic view of the planet from two different regions of the spectrum.

Because astronomers have long conjectured about the irregular dark spots discernible on the surface of Venus' atmosphere, data to resolve these questions would be of great scientific interest. If the spots were indeed breaks in the clouds, they would stand out with much better definition in the infrared spectrum. If the radiation came from the cloud tops, there would be no breaks and the temperatures at both frequencies measured by the infrared radiometer would follow essentially the same pattern.

The Venusian atmosphere is transparent to the 8-micron region of the spectrum except for clouds. In the 10-micron range, the lower atmosphere would be hidden by carbon dioxide. If cloud breaks existed, the 8-micron emissions would come from a much lower point, since the lower atmosphere is fairly transparent at this wavelength. If increasing temperatures were shown in this region, it might mean that some radiation was coming up from the surface.

As a result of the Mariner II mission, scientists have hypothecated that the cold cloud cover could be about 15 miles thick, with the lower base beginning about 45 miles above the surface, and the top occurring at 60 miles. In this case, the bottom of the cloud layer could be approximately 200 degrees F; at the top, the readings vary from about minus 30 degrees F in the center of the planet to temperatures of perhaps minus 60 degrees to minus 70 degrees F along the edges. This temperature gradient would verify the limb-darkening effect seen by the microwave radiometer.

At the center of Venus, the radiometer saw a thicker, brighter, hotter part of the cloud layer; at the limbs, it could not see so deeply and the colder upper layers were visible. Furthermore, the temperatures along the cloud tops were approximately equally distributed, indicating that both 8- and 10-micron "channels" penetrated to the same depth and that both were looking at thick, dense clouds quite opaque to infrared radiation.

Both channels detected a curious feature along the lower portion of the terminator, or the center line between the night and day sides of the planet. In that region, a spot was shown that was apparently about 20 degrees F colder than the rest of the cloud layer. Such an anomaly could result from higher or more opaque clouds, or from such an irregularity as a hidden surface feature. A mountain could force the clouds upward, thus cooling them further, but it would have to be extremely high.

The data allow scientists to deduce that not enough carbon dioxide was present above the clouds for appreciable absorption in the 10-micron region. This effect would seem to indicate that the clouds are thick and that there is little radiation coming up from the surface. And, if present, water vapor content might be 1/1,000 of that in the Earth's atmosphere.

Since the cloud base is apparently at a very high temperature, neither carbon dioxide nor water is likely to be present in quantity. Rather, the base of the clouds must contain some component that will condense in small quantities and not be spectroscopically detected.

As a result of the two radiometer experiments, the region below the clouds and the surface itself take on better definition. Certainly, heat-trapping of infrared radiation, or a "greenhouse" effect, must be expected to support the 800 degree F surface temperature estimated from the microwave radiometer data. Thus, a considerable amount of energy-blanketing carbon dioxide must be present below the cloud base. It is thought that some of the near-infrared sunlight might filter through the clouds in small amounts, so that the sky would not be entirely black, at least to human eyes, on the sunlit side of Venus. There also may be some very small content of oxygen below the clouds, and perhaps considerable amounts of nitrogen.

The atmospheric pressure on the surface might be very high, about 20 times the Earth's atmosphere or more (equivalent to about 600 inches of mercury, compared with our 30 inches). The surface, despite the high temperature, is not likely to be molten because of the roughness index seen in the earlier radar experiments, and other indicators. However, the possibility of small molten metal lakes cannot be totally ignored.

The dense, high-pressure atmosphere and the heat-capturing greenhouse effect could combine over long periods of time to carry the extremely high temperature around to the dark side of Venus, despite the slow rate of rotation, possibly accounting for the relatively uniform surface temperatures apparently found by Mariner II.

THE RADAR PROFILE: MEASUREMENTS FROM EARTH

In 1961, the Jet Propulsion Laboratory conducted a series of experiments from its Goldstone, California, DSIF Station, successfully bouncing radar signals off the planet Venus and receiving the return signal after it had travelled 70 million miles in 6½ minutes.

In order to complement the Mariner mission to Venus, the radar experiments were repeated from October to December, 1962 (during the Mariner mission), using improved equipment and refined techniques. As in 1961, the experiments were directed by W. K. Victor and R. Stevens.

The 1961 experiments used two 85-foot antennas, one transmitting 13 kilowatts of power at 2,388 megacycles, the other receiving the return signal after the round trip to and from Venus. The most important result was the refinement of the astronomical unit—the mean distance from the Earth to the Sun—to a value of 92,956,200 ±300 miles.

Around 1910, the astronomical unit, plotted by classical optical methods, was uncertain to 250,000 miles. Before the introduction of radar astronomy techniques such as those used at Goldstone, scientists believed that the astronomical unit was known to within 60,000 miles, but even this factor of uncertainty would be intolerable for planetary exploration.

In radar astronomy, the transit time of a radio signal moving at the speed of light (186,000 miles per second) is measured as it travels to a planet and back. In conjunction with the angular measurement techniques used by earlier investigators, this method permits a more precise calculation of the astronomical unit.

Optical and radar measurements of the astronomical unit differ by 50,000 miles. Further refinement of both techniques should lessen the discrepancy between the two values.

The 1961 tests also established that Venus rotates at a very slow rate, possibly keeping the same face toward the Sun at all times. The reflection coefficient of the planet was estimated at 12%, a bright value similar to that of the Earth and contrasted with the Moon's 2%. The average dielectric constant (conductivity factor) of the surface material seemed to be close to that of sand or dust, and the surface was reported to be rough at a wavelength of 6 inches.

The surface roughness was confirmed in 1962. Since it is known that a rough surface will scatter a signal, the radar tests were observed for such indications. Venus had a scattering effect on the radar waves similar to the Moon's, probably establishing the roughness of the Venusian surface as more or less similar to the lunar terrain.

Some of the most interesting work was done in reference to the rotation rate of Venus. A radar signal will spread in frequency on return from a target planet that is rotating and rough enough to reflect from a considerable area of its surface. The spread of 5 to 10 cycles per second noted on the Venus echo would suggest a very slow rotation rate, perhaps keeping the same face toward the Sun, or possibly even in a retrograde direction, opposite to the Earth's.

In the Goldstone 1962 experiments, Venus was in effect divided into observation zones and the doppler effect or change in the returned signal from these zones was studied. The rate of rotation was derived from three months of sampling with this radar mapping technique. Also, the clear, sharp tone characteristic of the transmitted radar signal was altered on return from Venus into a fuzzy, indistinct sound. This effect seemed to confirm the slow retrograde rotation (as compared with the Earth) indicated by the radar mapping and frequency change method.

In addition to these methods of deducing the slow rotation rate, two other phenomena seemed to verify it: a slowly fluctuating signal strength, and the apparent progression of a bright radar spot across from the center of Venus toward the outside edge.

As a result, JPL scientists revised their 1961 estimate of an equal Venusian day and year consisting of 225 Earth days. The new value for Venus' rotation rate around its axis is 230 Earth days plus or minus 40 to 50 days, and in a retrograde direction (opposite to synchronous or Sun-facing), assuming that Venus rotates on an axis perpendicular to the plane of its orbit.

Thus, on Venus the Sun would appear to rise in the west and cross to the east about once each Venusian year. If the period were exactly 225 days retrograde, the stars would remain stationary in the sky and Venus would always face a given star rather than the Sun.

A space traveller hovering several million miles directly above the Sun would thus see Venus as almost stopped in its rotation and possibly turning very slowly clockwise. All the other planets of our system including the Earth, rotate counterclockwise, except Uranus, whose axis is almost parallel to the plane of its orbit, making it seem to roll around the Sun on its side. The rotation direction of distant Pluto is unknown.

The Goldstone experiments also studied what is known as the Faraday rotation of the plane of polarization of a radio wave. The results indicated that the ionization and magnetic field around Venus are very low. These data tend to confirm those gathered by Mariner's experiments close to the planet.

The mass of Venus was another value that had never been precisely established. The mass of planetary bodies is determined by their gravitational effect on other bodies, such as satellites. Since Venus has no known natural satellite or moon, Mariner, approaching closely enough to "feel" its gravity, would provide the first opportunity for close measurement.

The distortion caused by Venus on Mariner's trajectory as the spacecraft passed the planet enabled scientists to calculate the mass with an error probability of 0.015%. The value arrived at is 0.81485 of the mass of the Earth, which is known to be approximately 13.173 septillion (13,173,000,000,000,000,000,000,000) pounds. Thus, the mass of Venus is approximately 10.729 septillion (10,729,408,500,000,000,000,000,000) pounds.

In addition to these measurements, the extremely precise tracking system used on Mariner proved the feasibility of long-range tracking in space, particularly in radial velocity, which was measured to within 1/10 of an inch per second at a distance of about 54 million miles.

As the mission progressed, the trajectory was corrected with respect to Venus to within 10 miles at encounter. An interesting result was the very precise location of the Goldstone and overseas tracking stations of the DSIF. Before Mariner II, these locations were known to within 100 yards. After all the data have been analyzed, these locations will be redefined or "relocated" to within an error of only 20 yards.

Mariner not only made the first successful journey to Venus—it also helped pinpoint spots in the Californian and Australian deserts and the South African veldt with an accuracy never before achieved.

CHAPTER 10
THE NEW LOOK OF VENUS

The historic mission of Mariner II to the near-vicinity of Venus and beyond has enabled scientists to revise many of their concepts of interplanetary space and the planet Venus.

The composite picture, taken from the six experiments aboard the spacecraft and the data from the DSIF radar experiments of 1961 and 1962 revealed the following:

- Interplanetary space between the Earth and Venus, at least as it was during the four months of Mariner's mission, had a cosmic dust density some ten-thousand times lower than the region immediately surrounding the Earth.

- During this period, the extremely tenuous, widely fluctuating solar winds streamed continually out from the Sun, at velocities ranging from 200 to 500 miles per second.

- An astronaut travelling through these regions in the last quarter of 1962 would not have been seriously affected by the cosmic and high-energy radiation from space and the Sun. He could easily have survived many times the amount of radiation detected by Mariner's instruments.

- The astronomical unit, as determined by radar, the yardstick of our solar system, stands at 92,956,200, plus or minus 300 miles.

- The mass of Venus in relation to the Earth's is 0.81485, with an error probability of 0.015%.

- The rotation rate of Venus is quite slow and is now estimated as equal to 230 Earth days, plus or minus 40 to 50 days. The rotation might be retrograde, clockwise with respect to a Sun-facing reference, with the Sun rising in the west and setting in the east approximately one Venusian year later. The planet seems to remain nearly star-fixed rather than permanently oriented with one face to the Sun.

- Venus has no magnetic field discernible at the 21,598-mile approach of Mariner II and at that altitude there were no regions of trapped high-energy particles or radiation belts, as there are near the Earth.

- The clouds of Venus are about 15 miles thick, extending from a base 45 miles above the surface to a top altitude of about 60 miles.

- At the resolution of the Mariner II infrared radiometer, there were no apparent breaks in the cloud cover. Cloud-top temperature readings are about minus 30 degrees F near the center (along the terminator), and ranging down to minus 60 degrees to minus 70 degrees F at the limbs, showing an apparent limb-darkening effect, which would indicate a hot surface and the absence of a supercharged ionosphere.

- A spot 20 degrees F colder than the surrounding area exists along the terminator in the southern hemisphere: a high mountain could exist in this region, but such an hypothesis is purely conjectural. A bright radar reflection is also found on the Equator in the same general region. Causes of these phenomena are not established.

- At their base, the clouds are about 200 degrees F and probably are comprised of condensed hydrocarbons held in oily suspension. Below the clouds, the atmosphere must be heavily charged with carbon dioxide, may contain slight traces of oxygen, and probably has a strong concentration of nitrogen.

- As determined by the microwave radiometer, Venus' surface temperature averages approximately 800 degrees F on both light and dark sides of the planet. Some roughness is indicated and the surface reflectivity is equivalent to that of dust and sand. No water could be present at the surface but there is some possibility of small lakes of molten metal of one type or another.

- Some reddish sunlight, in the filterable infrared spectrum, may find its way through the 15-mile-thick cloud cover, but the surface is probably very bleak.

- The heavy, dense atmosphere creates a surface pressure of some twenty times that found on the Earth, or equal to about 600 inches of mercury.

The mission was completed and the spacecraft had gone into an endless orbit around the Sun. But before Mariner II lost its sing-song voice, it produced 13 million data words of computer space lyrics to accompany the music of the spheres.

APPENDIX
SUBCONTRACTORS

Thirty-four subcontractors to JPL provided instruments and other hardware for Mariners I and II.

The subcontractors were:

Aeroflex Corporation Long Island City, New York	Jet vane actuators
American Electronics, Inc. Fullerton, California	Transformer-rectifiers for flight telecommunications
Ampex Corporation Instrumentation Division Redwood City, California	Tape recorders for ground telemetry and data handling equipment
Applied Development Corporation Monterey Park, California	Decommutators and teletype encoders for ground telemetry equipment
Astrodata, Inc. Anaheim, California	Time code translators, time code generators, and spacecraft signal simulators for ground telemetry equipment
Barnes Engineering Company Stamford, Connecticut	Infrared radiometers Planet simulator
Bell Aerospace Corporation Bell Aerosystems Division Cleveland, Ohio	Accelerometers and associated electronic modules
Computer Control Company, Inc. Framingham, Massachusetts	Data conditioning systems
Conax Corporation	Midcourse propulsion explosive valves

Buffalo, New York	Squibs
Consolidated Electrodynamics Corp. Pasadena, California	Oscillographs for data reduction
Consolidated Systems Corporation Monrovia, California	Scientific instruments Operational support equipment
Dynamics Instrumentation Company Monterey Park, California	Isolation amplifiers for telemetry Operational support equipment
Electric Storage Battery Company Missile Battery Division Raleigh, North Carolina	Spacecraft batteries
Electro-Optical Systems, Inc. Pasadena, California	Spacecraft power conversion equipment
Fargo Rubber Corporation Los Angeles, California	Midcourse propulsion fuel tank bladders
Glentronics, Inc. Glendora, California	Power supplies for data conditioning system
Groen Associates Sun Valley, California	Actuators for solar panels
Houston Fearless Corporation Torrance, California	Pin pullers
Kearfott Division General Precision, Inc. Los Angeles, California	Gyroscopes
Marshall Laboratories Torrance, California	Magnetometers and associated operational support equipment

Matrix Research and Development Corporation Nashua, New Hampshire	Power supplies for particle flux detectors
Menasco Manufacturing Company Burbank, California	Midcourse propulsion fuel tanks and nitrogen tanks
Midwestern Instruments Tulsa, Oklahoma	Oscillographs for data reduction
Mincom Division Minnesota Mining & Manufacturing Los Angeles, California	Tape recorders for ground telemetry and data handling equipment
Motorola, Inc. Military Electronics Division Scottsdale, Arizona	Spacecraft command subsystems, transponders, and associated operational support equipment
Nortronics Division of Northrop Corporation Palos Verdes Estates, California	Attitude control gyro electronic, autopilot electronic, and antenna servo electronic modules, long-range Earth sensors and Sun sensors
Ransom Research Division of Wyle Laboratories San Pedro, California	Verification and ground command modulation equipment
Rantec Corporation Calabasas, California	Transponder circulators and monitors
Ryan Aeronautical Company Aerospace Division San Diego, California	Solar panel structures
Spectrolab Division of Textron Electronics, Inc. North Hollywood, California	Solar cells and their installation and electrical connection on solar panels

State University of Iowa Iowa City, Iowa	Calibrated Geiger counters
Sterer Engineering & Manufacturing Company North Hollywood, California	Valves and regulators for midcourse propulsion and attitude control systems
Texas Instruments, Inc. Apparatus Division Dallas, Texas	Spacecraft data encoders and associated operational support equipment, ground telemetry demodulators
Trans-Sonic, Inc. Burlington, Massachusetts	Transducers

In addition to these subcontractors, over 1,000 other industrial firms contributed to the Mariner Project.

FOOTNOTES

[1] Throughout this book "Mariner" refers to the successful Mariner II Venus mission. Mariner I was launched earlier but was destroyed when the launch vehicle flew off course.

[2] For scientific reasons, distances from Venus are calculated from the center of the planet. Hereafter in this chapter, these distances will be reckoned from the surface.